T0136677

PEONIES *of the World*

POLYMORPHISM AND DIVERSITY

PEONIES *of the World*

POLYMORPHISM AND DIVERSITY

HONG De-Yuan

Kew Publishing
Royal Botanic Gardens, Kew

Missouri Botanical Garden
St Louis

First published in 2011 by
Royal Botanic Gardens, Kew,
Richmond, Surrey, TW9 3AB, UK
www.kew.org

ISBN 978-1-84246-458-8

British Library Cataloguing in Publication Data
A catalogue record for this book is available from the British Library

Production editor: Sharon Whitehead
Typesetting and page layout: Christine Beard
Publishing, Design & Photography, Royal Botanic Gardens, Kew

Cover design: Lyn Davies

Printed in the United Kingdom by by Henry Ling Limited

For information or to purchase all Kew titles please visit
www.kewbooks.com or email publishing@kew.org

Kew's mission is to inspire and deliver science-based plant conservation worldwide, enhancing the quality of life.

Kew receives half of its running costs from Government through the Department for Environment, Food and Rural Affairs (Defra). All other funding needed to support Kew's vital work comes from members, foundations, donors and commercial activities including book sales.

CONTENTS

PREFACE

This is the second book of the monograph series *Peonies of the World*, which will consist of three books. The first book, subtitled *Taxonomy and Phytogeography*, was published in 2010 by Kew and Missouri Botanical Garden. In that book, we made a comprehensive taxonomic revision of the genus *Paeonia* (Paeoniaceae), recognising 32 species and 26 subspecies. The third book in the series will be titled *Phylogeny and Evolution*.

This book contains part of the results of extensive fieldwork that I carried out across the whole distribution range of the genus *Paeonia*. This fieldwork was supported by five grants from the National Geographic Society. In the first book, I expressed my sincere thanks to the Society, and to the many colleagues and friends who assisted me in carrying out fieldwork. I thank them again here: without their help I could not have produced this series of books.

Here, I should express my sincere thanks to Drs ZHOU Shi-Liang, WANG Xiao-Quan, ZHANG Da-Ming and Luo Yi-bo and to retired Prof. PAN Kai-Yu, all in our laboratory, and to Dr RAO Guang-Yuan at Peking University, for their great help in my fieldwork. They are all associate authors of the first book. I am especially grateful to Anne Oveson who provided very warm hospitality and great help during our fieldwork in the Blue Mountains, Oregon, in 2005.

With regard to this book in particular, I should particularly express my cordial thanks to Dr Peter Raven, Emeritus President of Missouri Botanic Garden, and to my wife, PAN Kai-Yu, a retired professor of the Institute of Botany, the Chinese Academy of Sciences. Having seen the thousands of colour photos of natural peony populations I had taken during my fieldwork, they suggested (nearly simultaneously) that I prepare this second book. My sincere thanks are also due to Miss MA Li-Ming and Miss GONG Xiao-Lin, who helped me in the preparation of this book.

The present book contains 356 colour photographs, all taken at sites where peonies were growing naturally. The photographs cover all of the 33 species and 24 subspecies of *Paeonia* and are accompanied by concise descriptions of the morphology and distribution of each species. I do hope that they are sufficient to illustrate the polymorphism and diversity present in peonies.

It is worth noting here that this book lists 33 species of *Paeonia*, rather than the 32 species listed in *Taxonomy and Phytogeography*. This is because I recently raised the former *Paeonia decomposita* Hand.-Mazz. subsp. *rotundiloba* D. Y. Hong to specific status under the name *Paeonia rotundiloba* (D. Y. Hong) D. Y. Hong. The new species is distinguished from *P. decomposita* in having carpels mostly 3, less frequently 4 or 2, very occasionally 5 (rather than nearly always 5) and disk 8–15 mm high (rather than 4–9.6 mm high). In addition, there are differences in the number of leaf segments and in the shape of the terminal leaf segment. I regret that I did not pay enough attention to the former two characters when I prepared the first book.

INTRODUCTION

Polymorphism and diversity are two terms used to describe genetic variations between organisms. Polymorphism is a phenomenon in which two or more genotypes or states of a character coexist in an interbreeding population. It is a nearly universal phenomenon, occurring in almost all populations of sexually reproductive organisms. The term diversity has a much broader meaning and includes all kinds of genetic variations in populations, races, species, or even taxa above the species level. Diversity can be polymorphic (i.e. variation within an interbreeding population) or polytypic (i.e. variation between interbreeding populations), and therefore includes polymorphism. In this book, however, the emphasis is on polymorphism and the two terms are treated independently.

In the first book in this series *Taxonomy and Phytogeography* (Chapter 3), I published a comprehensive and detailed analysis of the morphological characters of plants of the genus *Paeonia* that was based on field observations, population sampling, critical examination of all available specimens and statistical analysis. This analysis helps us to understand the extent and nature of the polymorphism and diversity present within the genus. The colour photographs in the present book further illustrate the polymorphism and diversity within *Paeonia*. I hope that both the written analysis provided in *Taxonomy and Phytogeography* and the illustrations published in this book will be valuable to botanists, plant taxonomists and, in particular, to horticulturists. The polymorphism and diversity within *Paeonia* is briefly summarised below:

ROOTS

Plants of the genus *Paeonia* have one of three types of roots. Two root types can be recognised readily in the woody peonies (sect. *Moutan* DC.): most peony species have non-thickened roots that resemble those of most shrubs, whereas *P. delavayi* has thickened (tuberous) roots. All of the herbaceous peony species have thickened roots that are either carrot-shaped or tuberous (fusiform). Therefore, the root is polytypical in peonies, but is relatively stable within species.

STEMS

Two types of stems are found in peonies: tree peonies have lignified woody stems, whereas herbaceous peonies have softer stems.

LEAVES

The lowermost one or two leaves on the shoots or stems are the best developed and have the most leaflets. When describing leaves, I always refer to the leaves at this position because they can be compared between individuals, populations or species. Peony leaves are polytypical. For example, I have found nine types of compound leaves in peonies: ternate, biternate, triternate, ternate-pinnate, biternate-pinnate, triternate-pinnate, ternate-bipinnate, biternate-bipinnate and quartiternate-pinnate. The shape of the compound leaves is usually characteristic of a particular species, and can be used to distinguish species. Nevertheless, leaves are polymorphic in some cases. For example, in the *Paeonia corsica* population at Mt Cagna, S Corsica (*D. Y. Hong et al.* H01015), the number of leaflets and/or segments was usually 9, but individual leaves with 10, 12, 13, 16 or even 20 leaflets and/or segments were found. In the *P. delavayi* population in Dawu County, W Sichuan, China (*D. Y. Hong et al.* H95063), leaf segments varied in number from 68 to 312. The indumentum of leaves is also polymorphic in a number of species. For example, in the *P. corsica* population mentioned above (*D. Y. Hong et al.* H01015), the leaves were glabrous, nearly glabrous, or sparsely to rather densely villous on the lower surface.

NUMBER OF FLOWERS

In tree peonies (sect. *Moutan*), there are several flowers on a shoot forming a cyme in subsect. *Delavayanae*, whereas the flowers are solitary and terminal in subsect. *Vaginatae*. The flowers of herbaceous peonies are mostly solitary and only the four species of sect. *Paeonia* subsect. *Albiflorae* have several flowers per stem. This character can, however, be polymorphic, particularly in *P. anomala*. In the *P. anomala* population at Wuzliti, Habahe County, N Xinjiang, China (*T. H. Ying* 1006, *T. H. Ying* 1007, *T. H. Ying* 1009, *T. H. Ying* 1010 and *T. H. Ying* 1022) stems were found to have a single flower without any additional undeveloped flower bud, a single flower with one or two additional undeveloped flower buds, or two flowers.

SEPALS

Both the number and the shape of peony sepals are diverse, but these characters are usually uniform within species. For example, the sepals of the woody peony *Paeonia jishanensis* are all rounded at apex, whereas those of closely related species such as *P. qiui* and *P. rockii* are caudate. Polymorphism in sepal characters can be seen within some species. For example, the individuals with 4, 5 or 6 sepals were found in the *P. delavayi* population *D. Y. Hong et al.* H97095 in NW Yunnan; furthermore, green, dark red and purple colour variants were also found within this population.

PETALS

Petals show great diversity within the genus *Paeonia*. I have seen variation in petal number, from 4 to 13, as well as great variation in petal colour. Peony petals come in nearly every colour: white, yellow, pink, red, purple and dark purple. The petals can have brown red, red or dark purple blotches at their base on a background of various colours. Within some species, petal colour is uniform and characteristic. For example, the petals of *P. ludlowii* are always purely yellow, whereas those of *P. decomposita* and *P. rotundiloba* are always red. In many peony species, however, the petals are extremely polymorphic. For example, in the *P. delavayi* population *D. Y. Hong et al.* H97119 in Dêqên, NW Yunnan, China, the flowers varied in petal number, having 4, 5, 6, 7 or 10 petals. Two *P. delavayi* populations *D. Y. Hong et al.* H97112 and *D. Y. Hong et al.* H97128 in Shangri-la (Zhongdian) County, NW Yunnan, China are good examples of populations in which petal colour shows great variation, with individual plants having petals that are white, pale yellow, yellow with a red or brown-red blotch at the base, orange, red, purple, or dark purple.

CARPELS

Carpel number is an important character that is often used to distinguish plant families and that varies a great deal within *Paeonia*. Carpel number is characteristic for some species of peonies; for example, *P. ludlowii* has a single carpel or very rarely two carpels, whereas *P. delavayi* has mostly 2–4 carpels. In general, carpel number is greatly polymorphic in a number of *Paeonia* species. For example, carpels were found to vary in number continuously from 1 to 8 in the *P. corsica* population *D. Y. Hong et al.* H01015 on Mt Cagna, S Corsica. Carpels exhibit polymorphism not only in number, but also in indumentum and colour. In the *P. corsica* population mentioned above (*D. Y. Hong et al.* H01015), they were glabrous to holosericeous. In the *P. lactiflora* population on Mt Huanggangliang, Inner Mongolia (*D. Y. Hong et al.* H04040), carpel numbers varied from 2 to 5, carpel indumenta were glabrous, scarcely hairy or clearly hairy, and carpel colours included green, red, purple and dark purple.

Traits that show polymorphism within the genus *Paeonia* also include stamen number, and the colours of filaments, anthers and disks.

This analysis clearly shows that there can be numerous forms within a single population of peony plants. Their relationships are those between parents and offspring or between siblings, and therefore these numerous forms are just members of a single family.

Throughout the history of systematics, polymorphism has greatly influenced taxonomic treatments and, in many instances, a lack of knowledge and understanding of polymorphism has caused great taxonomic problems. Users of herbarium specimens or other taxonomic materials often feel a degree of uncertainty or difficulty in identifying the plants in their hands. Some of this difficulty is due to the fact that some or many of the materials that they are looking at are "paper species", found only in specimens and taxonomic publications but not in nature.

I believe that most paper species have been described and named by taxonomists who have no knowledge of population biology and have no population concept in their minds. Thus, they do not understand polymorphism. When polymorphism is not taken into account and taxonomists work with just a few herbarium specimens, it is possible to slip into describing a specimen showing one of numerous forms as a new species, and perhaps even a second specimen of a different form from within the same population as a second new species. For example, the tree peony in NW Yunnan and SW Sichuan has been given five different species names because of its polymorphism in flower colour and leaf-segment shape. A specimen with red petals was used as the type for *P. delavayi* Franch., whereas the type specimen of *P. lutea* Delavay ex Franch. has purely yellow petals, that of *P. trollioides* Stapf has golden-yellow petals, that of *P. delavayi* var. *alba* Bean has white petals, and that of *P. potaninii* Kom. has narrower leaf segments. All of these forms can, however, be found in single populations such as *D. Y. Hong et al.* H97112. Thus, *P. lutea* Delavay ex Franch., *P. trollioides* Stapf, *P. delavayi* var. *alba* Bean and *P. potaninii* Kom. are "paper species" or varieties of *P. delavayi* Franch. A reader who has a tree peony from NW Yunnan with yellow petals but with a red blotch at the base would wonder where he should go, to *P. delavayi* or to *P. lutea*?

Another even more outstanding example is the peony in Corsica and Sardinia, two islands in the Mediterranean. The peonies on these two islands have been assigned to eight different taxa (four species and four varieties), for which 30 different botanical names have been used. Our expedition to these two islands in 2001 and subsequent studies have unravelled the mystery of how many

Paeonia taxa can actually be distinguished. We found a peony population on Mt Cagna, S Corsica, *D. Y. Hong et al.* H01015, that was extremely polymorphic. The number of leaflets or segments found on plants from this population was usually 9, but individuals with 10, 12, 13, 16 or even 20 leaflets and/or segments were also found. Furthermore, these leaflets were either glabrous or sparsely to rather densely villose beneath. Carpel numbers varied continuously from 1 to 8, and carpel indumenta from glabrous to holosericeous. We also saw numerous forms with various combinations of different character states. A similar phenomenon was found on Mt Limbardo, N Sardinia.

Paeonia corsica Sieber ex Tausch (1828) was first described as new on the basis of a type collection that has glabrous carpels and nearly glabrous leaves. By contrast, *P. glabrescens* Jord. was based on a collection with nearly glabrous leaves, and *P. morisii* on specimens with 9, rarely 7–20, leaflets that are hairy on the lower surface and with tomentose carpels. But all of these forms (and many more) could be found on Mt Cagna. Readers will doubtless be astonished to see so many new taxa described for the peony from Corsica and Sardinia and so many botanic names used for it. A puzzle was created by taxonomists who did not realise the great polymorphism present among the peony plants on these two islands, and who thus described different elements of a family as different taxa, species or varieties, or used varying botanical names for them.

The outcome is completely different if a taxonomist keeps the population concept in mind and uses a proper methodology when conducting taxonomic studies. It is essential for a taxonomist to carry out character analysis that is based on field observations, population sampling and/or examination of all of the specimens available, as well as subsequent statistical analysis. After doing so, he should be able to evaluate characters and to indicate which characters are stable or polymorphic, and he should understand the patterns of variation and the range of characters that show variation. Hence, he should be able to indicate which characters can be used for the delimitation of taxa. At this stage of the study, a taxonomist should be able to describe a particular species with a precise circumscription and to clarify its relationships with its relatives. A key made as the final result of such a study should be functional and easily used. Hopefully, I achieved these goals in the first book in this series on peonies, *Taxonomy and Phytogeography*. I am grateful for Jo Bennison's comments on this first book published in the *Newsletter* of *Peony Society* (2010): "Taking this book down the rows of species growing in the field, I found I could name and confidently identify species grown from seed where other books had left doubts and question marks. Suddenly things made a lot more sense and the different groups became clear."

The polymorphism described in the present book could serve as an object lesson for taxonomists, particularly those who are unable to undertake extensive fieldwork, reminding us to always exercise caution when compiling a taxonomic treatment. The cases of polymorphism presented here may also be valuable examples for those who teach population biology and biodiversity.

Polymorphism and diversity are genetic resources that are very valuable for both plant breeders and horticulturists. The polymorphism and diversity in peonies described in the present book are remarkable, and thus the genus *Paeonia* has been shown to be rich in genetic resources. There are many forms (genotypes) of peony that are very attractive but have not been utilised for horticultural purposes. For example, *P. daurica* subsp. *mlokosewitschii* (from the Caucasus, Georgia) has a great variety of petal colour forms, one of which is white but with a pink to red periphery. I have not yet found this beautiful peony in gardens. There is a population of *P. lactiflora* in Mt Huanggangliang of Inner Mongolia in which the carpels exhibit great variations in colour and number that could greatly enrich the cultivated varieties of the famous herbaceous peony.

KEY TO SPECIES

1a. Shrubs; disk leathery and halfway to wholly enveloping carpels until mid-anthesis, or fleshy and short, enveloping only the base of carpels. . . . . I. sect. ***Moutan*** DC.

2a. Flowers usually 2–4 in a cyme, more or less pendent; disk fleshy, enveloping only the base of the carpels; carpels always glabrous . . . . Ia. subsect. ***Delavayanae*** Stern

3a. Carpels usually 2–5(–7); follicles less than 4 cm long, 1.5 cm in diameter; petals, filaments and stigmas often not purely yellow . . . . 2. *P. delavayi*

3b. Carpels nearly always single, rarely 2; follicles 4.7–7 cm long, 2–3.3 cm in diameter; petals, filaments and stigmas always yellow. . . . . .1. *P. ludlowii*

2b. Flowers solitary, erect; disk leathery, enveloping carpels halfway or completely until mid-anthesis; carpels tomentose or glabrous . . . . .Ib. subsect. ***Vaginatae*** Stern.

4a. Carpels glabrous, 2–5; disk enveloping carpels halfway or up to the base of the styles until mid-anthesis; lower leaves decompound with leaflets (29–)33–63 in number, all lobed

5a. Carpels almost always 5, rarely 4 or 3; disk enveloping carpels halfway at anthesis; leaflets 35–63 in number, elliptic to narrowly rhomboid . . . . . . 3. *P. decomposita*

5b. Carpels mostly 3, less often 4 or 2, rarely 5; disk enveloping carpels up to base of styles at anthesis; leaflets (19–)25–35(–39) in number, rhomboid to nearly orbicular . . . . 4. *P. rotundiloba*

4b. Carpels densely lanate or tomentose, 5(–7); disk completely enveloping carpels until mid-anthesis; lower leaves biternate, biternate-pinnate or ternate-bipinnate; leaflets usually fewer than 20(–33) in number, if more at least some of them entire

6a. Lower leaves biternate; leaflets 9 in number, very occasionally 11 or 15 in *P. jishanensis*

7a. Leaflets ovate or ovate-orbicular, mostly entire, often reddish above; petals often with a reddish blotch at the base. . . . . 8. *P. qiui*

7b. Leaflets oval, ovate or nearly orbicular, mostly or all lobed; green above; petals without a blotch at the base

8a. Leaflets oval or ovate, terminal leaflets 3- or 5-cleft, with additional 1 to several lobes, lateral leaflets mostly 2- or 3-lobed, less frequently entire; lobes acute at apex; leaves glabrous on lower surface; sepals all caudate or mucronate . . . . 9. *P. cathayana*

8b. Leaflets ovate-orbicular to orbicular, all 3-cleft; segments lobed, acute to rounded at apex; leaves villose along veins on lower surface; sepals all rounded at apex . . . . .7. *P. jishanensis*

6b. Lower leaves ternate-pinnate, ternate-bipinnate or biternate-pinnate; leaflets more than 9 in number, usually oval to lanceolate, mostly entire, less frequently oval-orbicular and mostly lobed

9a. Lower leaves ternate-pinnate; leaflets no more than 15 in number, ovate to ovate-lanceolate, mostly entire; petals white, rarely pale pink, without a blotch . . . . 6. *P. ostii*

9b. Lower leaves ternate-bipinnate (rarely biternate-pinnate); leaflets (17–)19–33 in number, lanceolate or ovate-lanceolate and mostly entire, or ovate to ovate-orbicular and mostly lobed; petals white, rarely red, always with a large, dark purple blotch at the base . . . . 5. *P. rockii*

1b. Herbaceous perennials; disk fleshy and short, enveloping only base of carpels

10a. Petals nearly equal in size to or smaller than sepals; disk dentate, almost interrupted; lower leaves ternate or biternate, with leaflets 3 or 9, lateral roots slightly fusiform . . . . II. sect. ***Onaepia*** Lindl.

11a. Lower leaves biternate; leaf segments and final lobes 55–110 in number, acute to rounded at apex; carpels usually 5, rarely 3, 4 or 6 in number; sepals exceeding petals. . . . . 10. *P. brownii*

11b. Lower leaves ternate; leaf segments and final lobes 30–78 in number, mostly acute at apex; carpels usually 3, rarely 2 or 4; sepals slightly smaller than, or subequal to, petals . . . . 11. *P. californica*

10b. Petals much larger than sepals; disk annular, waved or flat; lower leaves biternate or triternate, with leaflets 9 or more; lateral roots carrot-shaped, fusiform or tuberous . . . . III. sect. ***Paeonia***

12a. Flowers several to a stem, less frequently solitary but with 1–2 undeveloped flower buds at axils, rarely really solitary; sepals mostly caudate at apex; leaves usually with bristles along veins above. . . . . IIIa. subsect. ***Albiflorae*** (Salm-Dyck) D. Y. Hong

13a. Leaves cartilaginous and denticulate on margin, pubescent along veins or glabrous beneath; carpels glabrous, very rarely hairy. . . . . 12. *P. lactiflora*

13b. Leaves smooth on margin, glabrous beneath; carpels glabrous or hairy

14a. Leaflets or leaf segments of lower leaves 70–100 in number. . . . . . 15. *P. anomala*

14b. Leaflets or leaf segments of lower leaves fewer than 30 in number

15a. Carpels 1, rarely 2 in number, mostly tomentose, less frequently glabrous; flowers often several per stem . . . . 13. *P. emodi*

15b. Carpels mostly 2 or 3 in number, rarely 4, always glabrous; flowers nearly always solitary, rarely 2 in number, but sometimes with 1–2 undeveloped flower buds at axils. . . . . 14. *P. sterniana*

12b. Flowers always solitary and terminal; sepals mostly rounded at apex; leaves glabrous or with bristles along veins above

16a. Lateral roots always tuberous; leaflets or leaf segments of lower leaves mostly more than 20 in number, rarely fewer; leaves covered with bristles along veins above, or, if glabrous above, nearly always villose beneath . . . . . . IIIc. subsect. ***Paeonia***

17a. Stems mostly hirsute; sepals mostly hairy on abaxial side; leaves always glabrous or rarely villose at base above, more or less villose beneath

18a. Leaflets or leaf segments of lower leaves mostly more than 20, very rarely as few as 11 in number, linear-elliptic or lanceolate; sepals hispidulous or glabrous on abaxial side . . . . 33. *P. officinalis*

18b. Leaflets or leaf segments of lower leaves mostly less than 20, occasionally up to 32 in number; elliptic, oblong or ovate-lanceolate; sepals densely villose on abaxial side

19a. Flowers rose to red; anthers yellow; leaflets or leaf segments of lower leaves 11–25, occasionally up to 32 in number . . . . 31. *P. arietina*

19b. Flowers dark purple; anthers orange; leaflets or leaf segments of lower leaves 9–15, occasionally up to 25 in number . . . . . . 32. *P. parnassica*

17b. Stems glabrous; sepals always glabrous; leaves mostly with bristles along veins above

20a. Leaflets or leaf segments always dentate-lobed, with lobes less than 1 cm long; stigmas yellow or pale pink. 29. *P. peregrina*

20b. Leaflets or leaf segments entire or rarely deeply lobed; stigmas red

21a. Leaflets or leaf segments of lower leaves 19–45 in number; leaves hispidulous beneath. . . . . 30. *P. saueri*

21b Leaflets or leaf segments of lower leaves more than 70 in number; leaves always glabrous beneath

22a. Leaflets or leaf segments of lower leaves 130–340 in number, mostly filiform, 0.5–8 mm wide . . . . 28. *P. tenuifolia*

22b. Leaflets or leaf segments of lower leaves 70–100 in number, linear, 4–18 mm wide . . . . 27. *P. intermedia*

16b. All roots carrot-shaped; leaflets or leaf segments of lower leaves numbering up to 21 (only in *P. broteri* up to 32 and in *P. clusii* 23–95); leaves always glabrous above . . . . IIIb. subsect. ***Foliolatae*** Stern

23a. Leaflets or leaf segments of lower leaves 23–95 in number, linear to ovate . . . . .20. *P. clusii*

23b. Leaflets or leaf segments of lower leaves numbering less than 21 (up to 32 in *P. broteri*), broad elliptic to obovate

24a. Leaflets or leaf segments acuminate to caudate-acuminate at apex; carpels densely brown-papillose or hispidulous, rarely glabrous . . . . 23. *P. mairei*

24b. Leaflets or leaf segments acute to mucronate at apex; carpels tomentose or glabrous

25a. Lower leaves with 9 or fewer leaflets; carpels always glabrous

26a. Carpels 3–10, mostly 4–6 in number; plants purple-red and glabrous throughout . . . . 16. *P. cambessedesii*

26b. Carpels mostly 2 or 3, rarely 1, 4 or 5 in number; plants usually green; leaves more or less hirsute beneath . . . . 16. *P. obovata*

25b. Lower leaves usually with 10–15 leaflets or segments, less frequently 9 or more; carpels mostly tomentose

27a. Carpels 1, less frequently 2 in number, nearly always glabrous, very occasionally sparsely hairy; follicles columnar, 4–5.4 cm long . . . . 26. *P. algeriensis*

27b. Carpels (1)2–4 in number, rarely more, mostly tomentose; follicles long-ovoid or ellipsoid, up to 4 cm long

28a. Styles 1.5–3.5 mm long; leaflets or leaf segments of lower leaves 11–14, rarely up to 17 in number, densely or sparsely villose beneath; carpels always glabrous . . . . 24. *P. kesrouanensis*

28b. Styles usually absent (only in *P. corsica* 1.5–3 mm long); leaflets/leaf segments of lower leaves 9–20 in number, glabrous, sparsely hispid, puberulent or villose beneath; carpels tomentose or glabrous

29a. Carpels always lanate or tomentose with hairs 2–3 mm long; styles absent

30a. Leaflets usually entire, leaflets or leaf segments of lower leaves 9, rarely 10, very occasionally 11 in number, usually ovate, rounded, nearly truncate with a mucro, or rounded, less frequently acute at apex 21. *P. daurica*

30b. Leaflets at least some segmented, leaflets or leaf segments of lower leaves usually numbering 10 or more, very occasionally 9, usually acute at apex

31a. Leaflets or leaf segments of lower leaves mostly (11)15–21, rarely up to 32 in number, 4–10(–15) cm long, 1.5–5(–6.5) cm wide; leaves always glabrous; hairs on carpels 2 mm long. . . . . 19. *P. broteri*

31b. Leaflets or leaf segments of lower leaves mostly (9)11–15, rarely up to 21 in number, 9–18 cm long, 4.5–9 cm wide; leaves sparsely hispid or glabrous; hairs on carpels 3 mm long . . . . 22. *P. mascula*

29b. Carpels glabrous or tomentose with hairs 1.5 mm long; styles present or absent

32a. Carpels tomentose, rarely glabrous; leaflets or leaf segments of lower leaves usually 9, rarely up to 20 in number, holosericeous beneath; styles 1.5–3 mm long. . . . . 18. *P. corsica*

32b. Carpels glabrous, very occasionally sparsely hairy; leaflets or leaf segments of lower leaves 10–15 in number, glabrous, very occasionally sparsely hairy (hairy leaves always accompanied by hairy carpels); styles lacking. . . . . 25. *P. coriacea*

ILLUSTRATED DESCRIPTIONS OF TAXA

I. Sect. *Moutan* DC.

Shrubs to 3.5 m tall; stems always sympodial, the annual shoots dying back at the top. Flowers solitary or several per shoot, forming a cyme; sepals all caudate at the apex except in *P. jishanensis*; disk low and fleshy or raised and leathery, enveloping carpels halfway or entirely until mid-anthesis.

Subsect. *Delavayanae* Stern

Flowers several in a cyme. Floral disk fleshy, short, enveloping only the base of the carpels; carpels 1–5 in number.

1. *Paeonia ludlowii* (Stern & G. Taylor) D. Y. Hong (Figs 1.1–1.10)
2. *Paeonia delavayi* Franch. (Figs 2.1–2.31)

Subsect. *Vaginatae* Stern

Flowers solitary and terminal. Disk leathery, raised, enveloping carpels halfway or completely until mid-anthesis; carpels mostly 5 in number.

3. *Paeonia decomposita* Hand.–Mazz. (Figs 3.1–3.10)
4. *Paeonia rotundiloba* (D. Y. Hong) D. Y. Hong (Figs 4.1–4.5)
5. *Paeonia rockii* (S. G. Haw & Lauener) T. Hong & J. J. Li ex D. Y. Hong
    subsp. *rockii* (Figs 5.1–5.6)
    subsp. *atava* (Brühl) D. Y. Hong & K. Y. Pan (Figs 5.7–5.14)
6. *Paeonia ostii* T. Hong & J. X. Zhang (Figs 6.1–6.6)
7. *Paeonia jishanensis* T. Hong & W. Z. Zhao (Figs 7.1–7.10)
8. *Paeonia qiui* Y. L. Pei & D. Y. Hong (Figs 8.1–8.11)
9. *Paeonia cathayana* D. Y. Hong & K. Y. Pan (Figs 9.1–9.3)

1. *Paeonia ludlowii* (Stern & G. Taylor) D. Y. Hong, *Novon* 7(2): 157. Figs 1 and 2 (1997).
Paeonia lutea Delavay ex Franch. var. *ludlowii* Stern & G. Taylor

Deciduous and caespitose shrubs, up to 3.5 m tall, glabrous throughout. Roots attenuate downward, not fusiform. Stems up to 4 cm in diameter. Lower leaves biternate, green above, pale glaucous beneath; leaflets 9; leaflets 6–19 cm long, 5–15 cm wide, 3-segmented to half or nearly to the base; segments 4–12 cm long, 1.5–5.5 cm wide, mostly 3–lobed to the middle; lobes 2–5 cm long, 0.5–2.5 cm wide, segments, lobes, and teeth all acuminate at the apex. Flowers 3 or 4 on each shoot, both terminal and axillary, with the terminal one blooming first, forming a cyme; pedicels slightly curved, 5–14 cm long, naked or with a leafy bract. Involucrate bracts 4 or 5. Sepals 3 or 4, all or all except one caudate at the apex. Petals pure yellow, spreading, obovate, rounded at the apex, 4–5.5 cm long,

2.5–3.5 cm wide. Filaments yellow, anthers yellow. Disk fleshy, 1 mm high, yellow, waved. Carpels mostly single, very rarely 2; stigmas sessile, yellow. Follicles cylindrical, 4.7–7 cm long, 2–3.3 cm in diameter. Seeds kidney-shaped, dark brown, ca. 1.5 cm long, 1.2 cm in diameter.

Flowering from late May to early June; fruiting in September.

Chromosome number: 2n = 10.

Growing in sparse forests and thickets on granites, 2,870–3,450 m altitude.

Confined to SE Tibet: Nyingchi, Mailing and Lhünzê.

Fig. 1.1. Habitat and caespitose habit, 2,900 m, Gangga, Mailing County, SE Tibet, 24 May 1996. [Voucher: *D. Y. Hong, Y. B. Luo & S. R. Zhang* H96005 (A, K, MO, PE, US).]

Opposite:

Fig. 1.2. Habitat and habit: a plant 3.5 m tall, in the population at 2,980 m, Zhare Township, Mailing County, SE Tibet, 24 May 1996. [Voucher: *D. Y. Hong, Y. B. Luo & S. R. Zhang* H96007 (A, K, MO, PE, US).]

Fig. 1.3. A shoot, showing a cyme of three flowers, in the population shown in Fig. 1.1.

Fig. 1.4. A flower, showing purely yellow petals and stamens, and a single glabrous carpel, in the population shown in Fig. 1.1.

Fig. 1.5. A flower, showing purely yellow petals, involucrate bracts and sepals, in the population shown in Fig. 1.1.

Fig. 1.6. The back view of a flower showing involucrate bracts and sepals, also in the population shown in Fig. 1.1.

1.2
1.3
1.4
1.5
1.6

Fig. 1.7a–d. Disk and carpels showing variation in colour of disk and stigmas in the population shown in Fig. 1.1, but photographed on 17 May 2006.

Fig. 1.8. Two of the previous year's follicles, in the population shown in Fig. 1.1, but photographed on 17 May 2006.

Fig. 1.9. Four clumps from the population shown in Fig. 1.2 that have been dug for the use of peeling root bark as medicine. This usage is a serious threat to the survival of this rare species. The caespitose habit is clearly shown.

Fig. 1.10. Roots of a young individual in the population shown in Fig. 1.1, but photographed on 17 May 2006.

2. *Paeonia delavayi* Franch., *Bull. Soc. Bot. France* 33: 382 (1886).
Paeonia lutea Delavay ex Franch.; *P. potaninii* Kom.; *P. trollioides* Stapf ex Stern.

Shrubs 0.2–1.8 m tall, glabrous throughout. Roots tuberous with tubers up to 8 cm long, 2 cm in diameter. Shoots simple but often with sterile branches in axils. Lower 2 or 3 leaves biternate or ternate-pinnate, 15–30 cm long (excluding petioles), 10–22 cm wide; leaflets 9, first divided into 3–11 primary segments; segments divided again mostly to near the base or halfway into 2–11 secondary segments, thus each lower leaf with (17–)40–100(–312) segments; segments linear, linear-lanceolate, usually entire, 1.5–10 cm long, 0.5–4.5 cm wide; segments and lobes acute at the apex. Flowers usually 2–3 on a shoot, terminal and axillary, forming a cyme, less frequently solitary and terminal. Involucrate bracts 1–5, leaf-like. Sepals 2–9, green but pink at the base inside, or entirely purple or purple-red, rounded or triangular-rounded, mostly caudate, rarely rounded at the apex. Petals 4–13, but mostly 7–11, yellow, yellow with a red or purple-red spot at the base, red, dark red, or dark purple-red, sometimes white, orange, green-yellow, or yellow with red margins. Stamens 25–160; filaments yellow, pink, red, or dark purple-red; anthers yellow, pink, red, or dark purple-red; disk fleshy, 1–3 mm high, incised, green, yellowish, yellow, red, or dark red; carpels 2–4, very rarely 6–8. Ovaries usually green, sometimes purple; stigmas sessile, yellow-green, yellow, red, or purple-red. Follicles oblong-ovoid, 2–4 cm long, 1–1.5 cm wide, brown at maturity. Seeds 1–6 in each follicle, brown-black, oblong, ca. 10 mm long, ca. 8 mm in diameter.
Flowering from late April to mid June; fruiting from August to October.
Chromosome number: 2n = 10.
Growing at altitudes from 1,900 to 4,000 m, primarily in sparse thickets or dry woods, rarely on grassy slopes or glades of virgin *Picea* forests.
Endemic to SW China: W Sichuan, E Tibet and Yunnan.

Fig. 2.1. Habitat: in thickets of limestones, at 2,780 m, Liangwan Shan, Chenggong County, Yunnan, 24 May 1997. [Voucher: *D. Y. Hong et al.* H97078 (A, CAS, K, MO, PE, US).]

Fig. 2.2. Habitat: in *Picea* forest, at 3,200 m, the Yulong Snow Range, Lijiang County, Yunnan, 30 May 1997. [Voucher: *D. Y. Hong et al*. H97103 (A, CAS, K, MO, PE, US).]

Fig. 2.3. A large population in open woods, at 3,000 m, Zaba Village, between Nyingchi Town and Bayizhen, SE Tibet, 17 May 2006. [Voucher: *D. Y. Hong, Z. Q. Zhou & A. S. Xu* H06012 (PE).]

Fig. 2.4. An individual in the population shown in Fig. 2.3.

Fig. 2.5. A shoot showing a cyme of four flowers, also from the population shown in Fig. 2.3.

Fig. 2.6. Green-yellow petals in the population at 2,200 m, West Hill, Kunming, Yunnan, 23 May 1997. [Voucher: *D. Y. Hong et al*. H97077 (A, CAS, K, MO, PE, US).]

Fig. 2.7. Pure yellow petals in the population shown in Fig. 2.1.

Fig. 2.8. Yellow petals with a dark red-purple blotch at the base, in the population at 2,920 m, Mt Cangshan, Dali, Yunnan, 26 May 1997. [Voucher: *D. Y. Hong et al.* H97087 (A, CAS, K, MO, PE, US).]

Fig. 2.9. A population in stony *Quercus* thickets, showing the entirely herbaceous aerial part and petal colour, at 2,940 m, Ganghaizi, Lijiang, Yunnan, 30 May 1997. [Voucher: *D. Y. Hong et al.* H97095 (A, CAS, K, MO, PE, US).]

Fig. 2.10. Petals red but fringed with white, in the population shown in Fig. 2.9.

Fig. 2.11. Dark red petals, also in the population shown in Fig. 2.9.

Fig. 2.12. Dark red-purple petals, also in the population shown in Fig. 2.9.

Fig. 2.13. Back view of a flower showing involucrate bracts, sepals and yellow petals, from the population shown in Fig. 2.3.

Fig. 2.14. Variation in the colour and number of involucrate bracts and sepals, in the population shown in Fig. 2.9.

Fig. 2.15. Variation of carpels number (from 1 to 5) and colour, and of sepals colour, in the population shown in Fig. 2.9.

Fig. 2.16. A typical *P. delavayi* flower with 6 sepals and more involucrate bracts, in the population shown in Fig. 2.9.

Fig. 2.17. Two individuals (a single clone?) with red petals and entirely herbaceous aerial parts, in the population at 3,200–3,300 m, Hala Village, Zhongdian (Shangri-la), Yunnan, 3 June 1997. [Voucher: *D. Y. Hong et al.* H97112 (A, CAS, K, MO, PE, US).]

Fig. 2.18. Red petals in the population shown in Fig. 2.17.

Fig. 2.19. A flower in the population shown in Fig. 2.17, showing red petals, involucrate bracts and sepals.

Fig. 2.20. Orange petals in the population shown in Fig. 2.17.

Fig. 2.21. Petals yellow fringed with orange, in the population shown in Fig. 2.17.

Fig. 2.22. Pale yellow petals, also in the population shown in Fig. 2.17.

Fig. 2.23. Typical petal colour polymorphism, in the population shown in Fig. 2.17.

Fig. 2.24. Variation in the colour of sepals, disk, filaments, anthers and stigmas, in the population shown in Fig. 2.17.

Fig. 2.25. Yellowish white petals in the population at 3,040 m, Gezan Township, Zhongdian (Shangri-la), Yunnan, 6 June 1997. [Voucher: *D. Y. Hong et al.* H97128 (A, CAS, K, MO, PE, US).]

Fig. 2.26. An individual that has white petals from the same population as the individual shown in Fig. 2.25.

Fig. 2.27. Another petal colour polymorphism from the same population as the individual shown in Fig. 2.25.

Fig. 2.28. Disk of *P. delavayi*: **a**, a flower of the population *D. Y. Hong et al.* H06012, showing a yellow and dentate disk, purple-red filaments and three carpels; **b**, two flowers in the population at 2,600 m, in sparse pine woods, Guxiang Township, Bomi County, SE Tibet, showing a white and dentate disk, purple filaments, and two or three carpels, 15 May 2006. [Voucher: *D. Y. Hong, Z. Q. Zhou & A. S. Xu* H06015 (PE).]

Fig. 2.29. Follicles in the population at 2,900 m, Niri Township, Yajiang County, Sichuan, 24 August 1995. [Voucher: *D. Y. Hong, Y. B. Luo & Y. H. He* H95070 (A, CAS, K, MO, PE, US).]

Fig. 2.30. Roots fusiform-thickened, in the population shown in Fig. 2.17.

Fig. 2.31. Two types of reproduction in *P. delavayi*: **a**, seedling in the same population as that shown in Fig. 2.3; **b**, cloning by suckers, in the population shown in Fig. 2.17.

3. *Paeonia decomposita* Hand.-Mazz., *Acta Horti Gothob.* 13: 39 (1939).

Paeonia szechuanica W. P. Fang

Shrubs to 1.8 m tall, glabrous throughout. Stems up to 2 cm in diameter. Lower leaves mostly triternate-pinnate, ternate-bipinnate, biternate-bipinnate, with 29–65 leaflets; terminal leaflets elliptic to orbicular, 3-partite to the base or 3-fid, terminal lobes 3-lobed; lateral leaflets elliptic to orbicular, 3-lobed or coarsely toothed. Flowers solitary, terminal, 10–15 cm wide. Involucrate bracts 2–5, mostly 2 or 3, unequal, linear-lanceolate. Sepals 3–5, green, broadly obovate, all caudate at the apex. Petals 9–12, rose, obovate, 4–7 cm long, 3–5 cm wide. Disk leathery, enveloping half of carpels at anthesis, white or yellowish, with triangular teeth. Carpels always 5, very rarely 4 or 3, glabrous, green or purple; styles 1–2.5 mm long; stigmas red. Follicles black-brown when mature, ellipsoid, 2–4 cm long, 1.3–1.7cm in diameter. Seeds black, glossy, broadly ellipsoid or globose, 8–10 mm long, 6–8 mm in diameter.

Flowering in April and May; fruiting in August.

Chromosome number: 2n = 10.

Growing in young secondary deciduous broad-leaved forests, thickets, and sparse conifer forests, at an altitude of 2,050–3,100 m.

Endemic to NW Sichuan, China.

Opposite:

Fig. 3.1. Habitat: mountain slopes with thickets or sparse woods at 2,200–2,350 m in Dadu Valley, Zengda Village, Jinchuan County, Sichuan, 21 August 1995. [Voucher: *D. Y. Hong, Y. B. Luo & Y. H. He* H95037 (A, CAS, K, MO, PE, US).]

Fig. 3.2. Habitat: a population in thickets at 2,620 m in Zonggang Township, Barkam, Sichuan, 13 May 2006. [Voucher: *D. Y. Hong & Z. Q. Zhou* H06008 (PE).]

Fig. 3.3. A large clump in fruit at 2,400–2,500 m, between Hongqi Bridge and Guangyin Bridge, Jinchuan County, Sichuan, 20 August 1995. [Voucher: *D. Y. Hong, Y. B. Luo & Y. H. He* H95036 (A, CAS, K, MO, PE, US).]

Above:

Fig. 3.4. A population at the edges of *Betula–Picea* forests at 2,700 m, Adi Village, Barkam Town, Barkam, Sichuan, 13 May 2006. [Voucher: *D. Y. Hong & Z. Q. Zhou* H06010 (PE).]

Fig. 3.5. An individual in the population shown in Fig. 3.4.

Fig. 3.6. An individual with a rose-red flower and with a fruit of six follicles on the right, in the population shown in Fig. 3.4.

Fig. 3.7. An individual with a rose-red flower also in the population shown in Fig. 3.4.

Fig. 3.8. An individual with a red flower also in the population shown in Fig. 3.4.

Fig. 3.9a–d. Polymorphism of carpel colour in the population shown in Fig. 3.2.

Fig. 3.10. Follicles and shape of leaflets: **a**, an individual with five dehiscent follicles in the population featured in Fig. 3.3; **b**, an individual of the population at 2,700 m, Barkam Town, Barkam, Sichuan, 19 August 1995. [Voucher: *D. Y. Hong, Y. B. Luo & Y. H. He* H95035 (A, CAS, K, MO, PE, US)].

4. *Paeonia rotundiloba* (D. Y. Hong) D. Y. Hong, *J. Syst. Evol.* 49(5): 465 (2011).
Paeonia decomposita Hand.-Mazz. subsp. *rotundiloba* D. Y. Hong, *Kew Bull.* 52(4): 961, fig. 1A (1997).

Shrubs up to 2.5 m tall, 3 cm in diameter at the base, glabrous throughout. Stems grey-black. Lower leaves mostly biternate-pinnate or ternate-pinnate, with 19–39 leaflets; leaflets not decurrent; terminal leaflets rhomboid to orbicular, 2.1–5.5 cm long, 1.5–4.8 cm wide, 3-partite to the base or 3-fid, terminal lobes 3-lobed. Flowers solitary, terminal, 10–15 cm broad. Involucrate bracts 2–5, mostly 2 or 3, unequal in size, linear-lanceolate or broad-elliptic, lobed or segmented. Sepals 3–5, green, broadly obovate or nearly orbicular, unequal in size, 1.5–3 cm long, 1.5–2.5 cm wide, all caudate at the apex. Petals obovate or oblong, incised at the apex, 3.5–6.5 cm long, 2–4.6 cm wide. Disk leathery, pale yellow, enveloping carpels nearly to the base of style at anthesis, 8–15 mm high, with triangular teeth. Carpels mostly 3, less often 2 or 4, very rarely 5; styles 1–1.3 mm long; stigma red. Follicles brown or grey-brown when mature, ellipsoid, 2.2–3.5 cm long, 1.2–1.6 cm in diameter. Seeds black, glossy, broadly ellipsoid or nearly globose, 8–10 mm long, 6–8 mm in diameter.
Flowering in May; fruiting from late August to September.
Chromosome number: 2n = 10.
Growing in well-developed thickets, young secondary forests or sparse *Cupressus chengiana* forests, at an altitude of 1,700–2,700 m.
Confined to the Minjiang Valley of N Sichuan and Tewo County of SE Gansu, China.

Fig. 4.1. An individual with mature follicles of differing number (two, three and four) in *Corylus* thickets in the population at 1,750 m, near Maoxian Town, Maoxian County, Sichuan, 15 August 1995. [Voucher: *D. Y. Hong, Y. B. Luo & Y. H. He* H95015 (A, CAS, K, MO, PE, US).]

Fig. 4.2. The same individual as shown in Fig. 4.1.

Fig. 4.3. An individual showing two flowers and leaflets in the same populationas that shown in Fig. 4.1, 20 May 1996, by PAN Kai-Yu. [Voucher: *K. Y. Pan & Y. H. He* H96002 (A, K, MO, PE, US).]

Fig. 4.4. A flower showing a lacerate disk and three carpels in the population *K. Y. Pan & Y. H. He* H96002, also by PAN Kai-Yu.

Fig. 4.5. Cloning by suckers is occasionally found in the population at 2,500 m, Sergu Township, Heishui County, Sichuan, 16 August 1995. [Voucher: *D. Y. Hong, Y. B. Luo & Y. H. He* H95017 (A, CAS, K, MO, PE, US).]

5. *Paeonia rockii* (S. G. Haw & Lauener) T. Hong & J. J. Li ex D. Y. Hong, *Acta Phytotax. Sin.* 36 (6): 539 (1998).

Shrubs to 1.8 m tall. Roots cylindrical, attenuate downwards. Lower leaves ternate-pinnate, ternate-bipinnate or biternate-pinnate, leaflets 17–33, lanceolate to ovate-lanceolate and mostly entire, or ovate to ovate-orbicular and mostly lobed, 2–11 cm long, 1.5–4.5 cm wide, villose along veins beneath, truncate to cuneate at the base, acute or acuminate at the apex. Flowers solitary, terminal. Involucrate bracts 3–4, leaf-like. Sepals 4–7, ovate-lanceolate, ovate-orbicular or orbicular, all caudate at the apex. Petals 8–13, obovate, white or rarely red, always with a large and dark purple blotch at the base, 5–9 cm long, 4–7 cm wide. Filaments yellow; anthers yellow. Disk entirely enveloping carpels, pale yellow, leathery, dentate or lobed at the apex. Carpels 5, densely tomentose; styles 0.5–2 mm long, tomentose; stigmas pale yellow. Follicles oblong, densely yellow tomentose, ca. 3 cm long.
Flowering from late April to May; fruiting from late July to August.
Chromosome number: 2n = 10.
Growing usually in deciduous forests in limestone areas at an altitude of 850–2,800 m.
Endemic to China: SE Gansu, W Henan, W Hubei, Ningxia, Shaanxi and N Sichuan.
Two subspecies:

1a. Leaflets lanceolate to ovate-lanceolate, all or mostly entire . . . . 5a. subsp. *rockii*
1b. Leaflets ovate to ovate-orbicular, all or mostly lobed. . . . . 5b. subsp. *atava*

5a. *Paeonia rockii* subsp. *rockii* (Figs 5.1–5.6).

Distributed in SE Gansu, W Henan, W Hubei, S Shaanxi and N Sichuan.

5b. *Paeonia rockii* subsp. *atava* (Brühl) D. Y. Hong & K. Y. Pan, *Acta Phytotax. Sin.* 43: 175 (2005).
Paeonia moutan Sims subsp. *atava* Brühl; *P. rockii* subsp. *taibaishanica* D. Y. Hong (Figs 5.7–5.14).

Distributed in Ningxia and C and N Shaanxi.

Fig. 5.1. An individual with a flower bud in secondary deciduous forest, in the population at 1,100 m, Mt Mudanduo, Baotianmen Nature Reserve, Neixiang County, Henan, 30 April 1997. [Voucher: *D. Y. Hong, Y. Z. Ye & Y. X. Feng* H97015 (MO, PE).]

Fig. 5.2. A large individual near a house at bottom of Mt Mudanduo, introduced from the population shown in Fig. 5.1, 30 April 1997. [Voucher: *D. Y. Hong, Y. Z. Ye & Y. X. Feng* H97016 (A, CAS, K, MO, PE, US).]

Fig. 5.3. A flower from the plant shown in Fig. 5.2, with white petals that have a large dark-purple blotch at the base, typical of this species.

Fig. 5.4. The same flower as shown in Fig. 5.3, back view.

Fig. 5.5. A flower with pinkish white petals, in the population at 1,360 m, Zhangjiapo, Houping Township, Baokang County, Hubei, 5 May 1997. [Voucher: *D. Y. Hong, Y. Z. Ye & Y. X. Feng* H97051 (A, CAS, K, MO, PE, US).]

Fig. 5.6. A large individual in Zhangjiapo, Houping Township, introduced from the same population as the flower shown in Fig. 5.5, 5 May 1997, the researcher is my former PhD student, Yu-Xin FENG, now in St. Louis.

Fig. 5.7. A large individual in deciduous broad-leaved forest in the population at 1,750 m, Mt Taibai, Shaanxi, 24 May 1985. [Voucher: *D. Y. Hong & X. Y. Zhu* PB85061 (PE).]

Fig. 5.8. A flower of the individual shown in Fig. 5.7, showing white petals with a large dark-purple blotch at the base, consistent with those in the typical subspecies.

Fig. 5.9. A tree peony in a large population at 1,320 m, Longbagou, Xiaoshiwan Township, Ganquan County, Shaanxi, 4 May 2006. [Voucher: *D. Y. Hong, K. Y. Pan & Y. Ren* H06003 (A, BM, K, MO, PE).]

Fig. 5.10. A flower of the individual plant shown in Fig. 5.9.

Fig. 5.11. The flower shown in Fig. 5.10, showing white petals with a dark purple blotch at the base.

Fig. 5.12. An individual with red flowers in the same population as that shown in Fig. 5.9.

Fig. 5.13. A flower on the plant shown in Fig. 5.12, showing the nearly black blotches at the base of the petals.

Fig. 5.14. Another flower from a plant in the population shown in Fig. 5.9, with rose-red petals, back view.

6. *Paeonia ostii* T. Hong & J. X. Zhang, *Bull. Bot. Res. Harbin* 12 (3): 223, fig. 1 (1992).

Shrubs up to 1.5 m tall. Lower leaves ternate-pinnate; leaflets11–15, lanceolate to ovate-lanceolate, mostly entire, often terminal ones 2- or 3-lobed, rounded at the base, acute to acuminate at the apex, 5–13 cm long, 2.5–6 cm wide, glabrous on both surfaces but sometimes pubescent at the base or the lower part of major veins above. Flowers solitary, terminal, single. Involucrate bracts 3–6, leaf-like. Sepals 4–6, green-yellow, broad-elliptic or ovate-orbicular, shortly caudate or acute at the apex. Petals usually 11-14, white, rarely pinkish, obovate, 5.5–8 cm long, 4–6 cm wide. Filaments purple-red; anthers yellow. Disk entirely enveloping carpels, purple-red, leathery, dentate or lobed at the apex. Carpels 5, densely tomentose; stigmas sessile, red. Follicles oblong, densely brown-yellow tomentose. Seeds brown-black, oblong-spherical or spherical, 8–9 mm long, 7–8 mm in diameter.
Flowering in April to May; fruiting in August.
Chromosome number: 2n = 10.
Growing in deciduous broad-leaved forests or thickets at an altitude below 1,600 m.
Native to Anhui (Chaohu) and W Henan (Lushi County and Xixia County); widely cultivated in China as a traditional medicine.

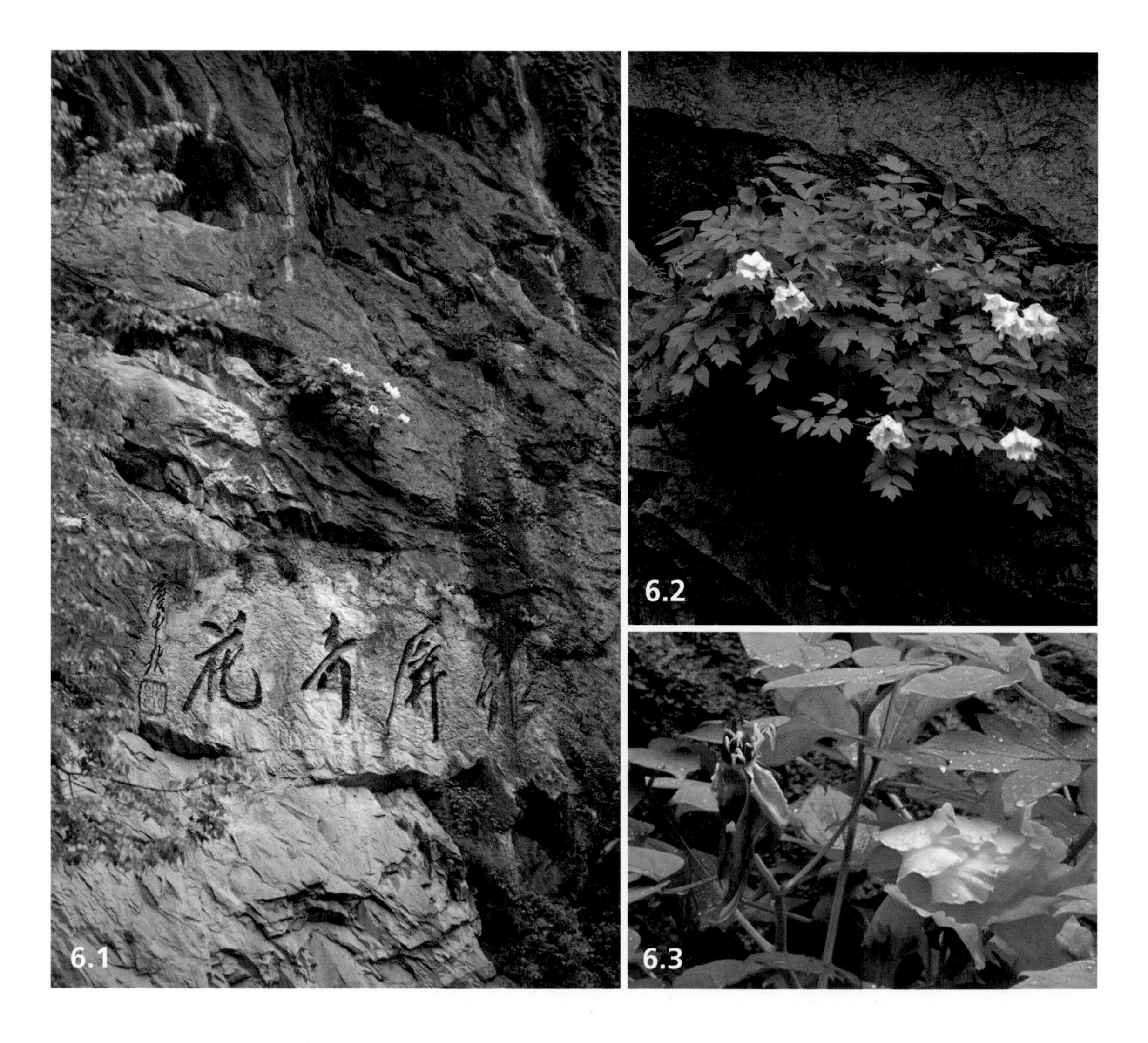

Opposite:

Fig. 6.1. An individual plant found ca. 40 m above the ground on a cliff of over 150 m high, Yinping Shan, Caohu, Anhui, 16 April 2006. The four Chinese characters in the cliff mean "Yinping Rare and Precious Flower" [Voucher: *K.Y. Pan & Z. W. Xie* 9701 (PE).]

Fig. 6.2. A close view of the plant shown in Fig. 6.1.

Fig. 6.3. A flower of the plant shown in Fig. 6.1 with petals shed, showing purple filaments and a purple disk. Photograph by Mr LI Min.

Above:

Fig. 6.4. A plantation of *P. ostii* grown for root bark ("dan pi"), a famous Chinese medicine, 29 April 1997.

Fig. 6.5. A flower with pinkish-white petals in the population at 1,400 m, Caijiacun, Guanpo Township, Lushi County, Henan, 16 May 1998. [Voucher: *D. Y. Hong et al*. H98005 (PE).]

Fig. 6.6. Another flower in the population shown in Fig. 6.5, showing narrow pinkish-white petals, purple filaments, yellow anthers, a purple disk and five carpels with red stigmas.

7. *Paeonia jishanensis* T. Hong & W. Z. Zhao, *Bull. Bot. Res. Harbin* 12 (3): 225, fig. 2 (1992).
Paeonia spontanea (Rehder) T. Hong & W. Z. Zhao; *P. suffruticosa* Andrews subsp. *spontanea* (Rehder) S. G. Haw & Lauener; *P. suffruticosa* Andrews var. *spontanea* Rehder.

Shrubs to 1.8 m tall. Roots attenuate downwards. Suckers present. Lower leaves biternate, with 9 leaflets, very occasionally terminal leaflets 3-fid to the base, and with 11 or 15 leaflets; leaflets ovate-orbicular or orbicular, 3-cleft, 4–8 cm long, 3–11 cm wide, glabrous above, villose along veins or throughout beneath; segments lobed, segments and/or lobes acute to rounded at the apex. Flowers solitary and terminal. Involucrate bracts 2–4, long-elliptic. Sepals 3 or 4, broad-ovate, all rounded at the apex. Petals 5–11 in number, white, occasionally pinkish at the base or on margins, obovate, 4.5–7.2 cm long, 4–6 cm wide. Filaments pink or purple, white above; anthers yellow. Disk entirely enveloping carpels at anthesis, red-purple, leathery, dentate at the apex. Carpels 5, densely tomentose; stigmas red. Follicles oblong, densely brown-yellow tomentose. Seeds dark brown, nearly spherical, 8–9 mm in diameter.
Flowering from April to May; fruiting in August.
Chromosome number: 2n = 10.
Growing in secondary deciduous broad-leaved forests or well-developed thickets at an altitude of 900–1,700 m.
Native to N Henan, C Shaanxi, and SW Shanxi.

Fig. 7.1. Habitat: a young secondary deciduous forest at 1,550 m, Majiagou, Xishe, Jishan County, Shanxi, 3 May 1993. [Voucher: *Y. L. Pei & D. Y. Hong* 93003 (PE).]

Fig. 7.2. Habitat: two individuals in a forest, at 1,650 m, Shuiyukoucun, Yongji County, Shanxi, 8 May 1993. [Voucher: *Y. L. Pei & D. Y. Hong* 93011 (PE).]

Fig. 7.3. Plant from the population shown in Fig. 7.2.

Fig. 7.4. Side view of a flower from the population shown in Fig. 7.2.

Fig. 7.5. Flower from the population shown in Fig. 7.2.

Fig. 7.6. A flower from the population shown in Fig. 7.2 with white petals that have a longitudinal pink stripe.

Fig. 7.7. The flower shown in Fig. 7.6, showing large yellowish sepals that have a rounded apex.

Fig. 7.8. A flower from the population shown in Fig. 7.2, showing pure white petals, purple filaments, yellow anthers, and a purple disk that entirely envelopes the five carpels each with a red stigma, 8 May 1993.

Fig. 7.9. A flower from the population shown in Fig. 7.2, showing a petaloid stamen.

Fig. 7.10. Cloning by suckers in the population shown in Fig. 7.2, a common phenomenon in this species.

8. ***Paeonia qiui*** Y. L. Pei & D. Y. Hong, *Acta Phytotax. Sin.* 33 (1): 91, fig. 1 (1995).

Shrubs up to 1.2 m tall. Roots up to 2 cm in diameter, attenuate downwards. Lower leaves biternate; leaflets always 9, often reddish above, ovate, ovate-lanceolate or ovate-orbicular, rounded at the base, obtuse or acute at the apex, mostly entire, sometimes terminal ones shallowly 3-lobed, 4–12 cm long, 2–8 cm wide, usually glabrous above, densely villose at axils of major veins beneath. Flowers solitary, terminal. Involucrate bracts 2–4, leaf-like. Sepals mostly 3, rarely 2 or 4, yellow green, acute or caudate at the apex. Petals 5–9, spreading, pink or whitish pink, often with a pale red spot at the base, 3.5–5.5 cm long, 2–3.1 cm wide. Filaments pale pink to pink; anthers yellow. Disk entirely enveloping carpels, red-purple, leathery. Carpels 5, densely tomentose; stigmas sessile, red, 1.5–2 mm wide. Follicles ellipsoid, densely brown-yellow tomentose, 2–2.8 cm long. Seeds black, glossy, 6–8 mm long, 5–7 mm in diameter.

Flowering from late April to May; fruiting in August.

Chromosome number: 2n = 10.

Growing mostly in secondary deciduous broad-leaved forests, rarely on sunny grassy slopes, on limestone rocks or cliffs, at an altitude of 1,000–2,200 m.

Confined to W Henan and W Hubei of China.

Opposite:

Fig. 8.1. Habitat: a peony on a limestone cliff at 1,300 m, Mt Chefengping, Houping Township, Baokang County, Hubei, 3 May 1997. [Voucher: *D. Y. Hong, Y. Z. Ye & Y. X. Feng* H97029 (A, K, MO, PE, US).]

Fig. 8.2. Habitat: an individual on a limestone cliff in a deciduous broad-leaved forest, in the population shown in Fig. 8.1.

Fig. 8.3. An individual in the population shown in Fig. 8.1.

Fig. 8.4. A flower from the population shown in Fig. 8.1.

Fig. 8.5. The flower shown in Fig. 8.4.

8.1
8.3
8.2
8.4
8.5

Fig. 8.6. An individual in a farm yard in Hongjiayuan Village, Houping Township, introduced from a nearby mountain, 2 May 1997. [Voucher: *D. Y. Hong, Y. Z. Ye & Y. X. Feng* H97023 (A, K, MO, PE, US).]

Fig. 8.7. An individual in the population at 1,800–2,010 m above Songbai Town, Shennongjia, Hubei, 25 April 1988. Photograph by Dr J. Z. Qiu. [Voucher: *J. Z. Qiu* PB88018 (PE).]

Fig. 8.8. A flower in the population shown in Fig. 8.7, showing a petaloid stamen. Photograph by Dr J. Z. Qiu.

Fig. 8.9. An individual on a limestone cliff in a deciduous forest on Mt Shantongya, above Songbai Town, Shennongjia, Hubei, 7 August 2004. [Voucher: *D. Y. Hong & Z. Q. Zhou* H04041 (PE).]

Fig. 8.10. A nearly mature follicle in the population shown in Fig. 8.9 (only one of the five carpels is developed).

Fig. 8.11. Roots from a plant in the population featured in Fig. 8.9.

9. ***Paeonia cathayana*** D. Y. Hong & K. Y. Pan, *Acta Phytotax. Sin.* 45 (3): 286, fig. 2 (2007).

Shrubs about 0.8 m tall. Leaves glabrous; lower leaves always biternate, and thus with 9 leaflets; terminal leaflets obovate-deltoid, 8–10 cm long, 7–9 cm broad, 3- or 5-cleft to middle or even beyond, lateral leaflets ovate or ovate-lanceolate, 4–7 cm long, 2–4.5 cm broad, entire or shallowly lobed. Flowers solitary, terminal, single. Involucrate bracts 2–6, glabrous. Sepals 4–5, all caudate at the apex, 3–3.5 cm long, 2–3 cm broad, glabrous. Petals 9 or 10, rose, broadly obovate, rounded at the apex, 5–6 cm long, 4–6 cm broad. Filaments purple, anthers yellow. Disk entirely enveloping the carpels at anthesis, purple; stigma purple.
Flowering from late April to early May.
Chromosome number: 2n = 10 (inferred; all the cultivars cytologically observed have 2n = 10, and thus the progenitor of these cultivars should be diploid).
Native to W Henan (Songxian) and W Hubei (Baokang).

Fig. 9.1. An individual at the side of YANG Hui-Fang's house, Muzhijie Township, Songxian County, Henan, introduced from a nearby mountain in the early 1960s (according to Mr Yang), 28 April 1997. [Voucher: *D. Y. Hong, Y. Z. Ye & Y. X. Feng* H97010 (PE).]

Fig. 9.2. The individual shown in Fig. 9.1.

Fig. 9.3. The individual shown in Fig. 9.1.

II. Sect. *Onaepia* Lindl.

Perennials. Lateral roots slightly fusiform. Stems often branched, lower leaves ternate or biternate. Flowers often several, terminal on main stem and branches; petals nearly equal in size to, or smaller than, sepals; disk fleshy, dentate, almost interrupted, enveloping the base of carpels until mid-anthesis; carpels 2–5 in number, always glabrous.

10. *Paeonia brownii* Douglas ex Hook. (Figs 10.1–10.8)

11. *Paeonia californica* Nutt. ex Torr. & A. Gray (Figs 11.1–11.9)

10. *Paeonia brownii* Douglas ex Hook., *Fl. bor.-amer.* 1: 27 (1829).

Perennial herbs glabrous throughout except leaf margins. Roots all slightly fusiform, up to 2.4 cm in diameter. Caudex (rhizomes) up to 12 cm long. Stems 15–48 cm tall, up to 1 cm in diameter, usually with no branches, but sometimes with fertile or sterile branches. Lower leaves biternate, with 9 leaflets; each leaflet with several segments, each segment with several final lobes; segments 0.3–2.0 cm wide; final lobes 59–110 in total, oblong or ovate-lanceolate, rounded or acute, sometimes mucronate at the apex, 0.2–1.2 cm wide, margins thickened and recurved, sometimes with bristles. Flowers terminal, solitary or up to 4 on a stem, pendent. Involucrate bracts 1–2, leaf-like. Sepals 3–5, green or purple or green but purple at the periphery, more or less larger than petals, rounded, 1–2.2 cm long, 1.1–2.1 cm wide. Petals 7–10, orbicular, red brown or brown-purple, often yellow at the periphery, entire, 0.8–1.5 cm long, 0.6–0.9 cm wide, incurved. Filaments pink; anthers yellow. Disk fleshy, dentate, 3 mm high. Carpels mostly 5, rarely 4; stigmas sessile, purple. Follicles cylindrical, 2–4 cm long, 1.2–1.9 cm in diameter. Seeds black, oblong, 10–12 mm long, 6–6.5 mm in diameter. Flowering from March to June, but mostly in May and early June.
Chromosome number: 2n = 10.
Growing usually in sparse chaparral, open places in woods, or on grassy slopes.
Distributed in western part of the USA: N California, Idaho, Nevada, Oregon, Utah, Washington and Wyoming.

Opposite:

Fig. 10.1. A population in clearings of conifer forests at 1,220 m, NW of Elgin, Union County, Blue Mts, Oregon, 24 May 2005. [Voucher: *D. Y. Hong, K. Y. Pan & P. Woodward* H05016 (A, BM, CAS, K, MO, PE).]

Fig. 10.2. An individual in the population shown in Fig. 10.1.

Fig. 10.3. A flower in the population 3 km N of Wallowa Town, Wallowa County, Blue Mts, Oregon, 25 May 2005. [Voucher: *D. Y. Hong, K. Y. Pan & P. Woodward* H05019 (A, BM, K, MO, PE).]

Fig. 10.4. Back view of a flower in the population shown in Fig. 10.1, showing two involucrate bracts and six sepals.

Fig. 10.5. A flower in the population shown in Fig. 10.1, showing a dentate and disrupted disk.

Fig. 10.6. Two flowers with young follicles from the population shown in Fig. 10.3, showing sepals of different colours.

Fig. 10.7. Involucrate bracts and sepals from a single flower in the population shown in Fig. 10.1, the innermost element showing a transition from sepal to petal.

Fig. 10.8. Underground parts: **a**, fusiform roots from the population shown in Fig. 10.1; **b**, a rhizome with two shoots and fusiform roots from the population at 1,930 m at Granny View Point, Imnaha, Wallowa County, Blue Mts, Oregon, 25 May 2005. [Voucher: *D. Y. Hong, K. Y. Pan & P. Woodward* H05023 (BM, K, MO, PE).]

10.1
10.2
10.3
10.4
10.5
10.6
10.8a
10.7
10.8b

11. *Paeonia californica* Nutt. ex Torr. & A. Gray, *Fl. N. Amer.* 1: 41 (1838).

Paeonia brownii subsp. *californica* (Nutt. ex Torr. & A. Gray) Halda

Perennials totally glabrous. Roots slightly fusiform-thickened, up to 3 cm in diameter. Caudex (rhizomes) up to 8 cm long. Stems 40–70 cm tall, up to 1.2 cm in diameter, usually with sterile or fertile branches. Lower leaves ternate, very occasionally nearly biternate; leaflets 3, each leaflet with several segments, each segment usually with two or several lobes, segments and/or lobes 33–78 in total; segments 3–8 cm long, 0.4–2.0 cm wide; lobes linear to lanceolate, usually acute, sometimes mucronate at the apex, 0.2–3.0 cm long, 0.2–1.2 cm wide. Flowers usually several, up to 6 on a stem, pendent. Involucrate bracts usually 1 or 2, leaf-like. Sepals 3 or 4, rounded, green or green but purple at the periphery, or purple, 1.2–2.0 cm long, 1.2–2.0 cm wide, as large as, or slightly smaller than petals. Petals 6–8, entire, purple, dull dark red, dark purple red, or brownish purple, 1.2–2.2 cm long, 1.0–2.0 cm wide. Filaments yellow; anthers yellow. Disk fleshy, yellow, dentate, teeth variable in shape, 2.5–6 mm high. Carpels 3, less frequently 2; stigmas sessile, 3 mm long, 2 mm wide, horizontal. Follicles cylindrical, 2.5–4.2 cm long, 1.1–1.7 cm in diameter. Seeds oblong, black, 10–12 mm long, 5.5–6 mm in diameter.

Flowering from January to May, but mostly in March and April; fruiting from June to July.

Chromosome number: 2n = 10.

Growing mostly in chaparral, or in openings or at the edges of chaparral and oak woods, from coastal areas at 30 m in altitude to mountain areas up to 1,200 m.

Distributed in the northern-most Baja California of Mexico and S California of the USA.

Fig. 11.1. A population among tall herbs with sparse shrubs at ca. 900 m, Angeles National Forest, Los Angeles County, California, 22 May 2005. [Voucher: *D. Y. Hong, K. Y. Pan & J. Z. Qiu* H05012 (A, BM, CAS, K, MO, PE).]

Fig. 11.2. A population in chaparral, San Bernardino National Forest, San Barnardino County, California, 21 May 2005. [Voucher: *D. Y. Hong, K. Y. Pan & J. Z. Qiu* H05011 (A, BM, CAS, K, MO, PE).]

Fig. 11.3. Several individuals at the edges of chaparral in the population shown in Fig. 11.2, March 2006. Photograph by Dr J. Z. Qiu.

Fig. 11.4. The population shown in Fig. 11.2.

Fig. 11.5. A flowering individual in the population shown in Fig. 11.2, March 2006. Photograph by Dr J. Z. Qiu.

Fig. 11.6. Back view of two flowers and leaves in the population shown in Fig. 11.2, March 2006. Photograph by Dr J. Z. Qiu.

Fig. 11.7. Two flowers with yellow dentate and disrupted disks from the population shown in Fig. 11.2.

Fig. 11.8. A flower with three young follicles from the population shown in Fig. 11.2.

Fig. 11.9. Slightly fusiform roots in the population shown in Fig. 11.2.

III. Sect. *Paeonia*

Perennials. Lateral roots carrot-shaped, fusiform or tuberous. Lower leaves biternate or triternate. Flowers solitary and terminal or several on a stem; sepals caudate or rounded at the apex; disk annular, fleshy, enveloping only the base of carpels until mid-anthesis; carpels 1–10 in number, glabrous or lanate, tomentose, hispidulous or papillose.

Subsect. *Albiflorae* (Salm-Dyck) D. Y. Hong

Roots more-or-less carrot-shaped. Leaves glabrous or with bristles along veins on the upper surface. Flowers usually several per stem, rarely solitary, or solitary but with undeveloped (sterile) flower buds at axils; sepals mostly caudate at the apex.

12. *Paeonia lactiflora* Pall. (Figs 12.1–12.9)
13. *Paeonia emodi* Wall. ex Royle (Figs 13.1–13.7)
14. *Paeonia sterniana* H. R. Fletcher (Figs 14.1–14.10)
15. *Paeonia anomala* L.
    - subsp. *anomala* (Figs 15.1–15.5)
    - subsp. *veitchii* (Lynch) D. Y. Hong & K. Y. Pan (Figs 15.6–15.17)

Subsect. *Foliolatae* Stern

Roots carrot-shaped. Lower leaves biternate; leaflets or leaf segments usually numbering 9 or more but fewer than 21 (up to 32 in *P. broteri* and 95 in *P. clusii*); leaves always glabrous above. Flowers always solitary and terminal; sepals mostly rounded at the apex.

16. *Paeonia obovata* Maxim.
    - subsp. *obovata* (Figs 16.1–16.16)
    - subsp. *willmottiae* (Stapf) D. Y. Hong & K. Y. Pan (Figs 16.17–16.22)
17. *Paeonia cambessedesii* (Willk.) Willk. (Figs 17.1–17.7)
18. *Paeonia corsica* Sieber ex Tausch (Figs 18.1–18.11)
19. *Paeonia broteri* Boiss. & Reut. (Figs 19.1–19.9)
20. *Paeonia clusii* Stern
    - subsp. *clusii* (Figs 20.1–20.5)
    - subsp. *rhodia* (Stearn) Tzanoud. (Figs 20.6–20.8)
21. *Paeonia daurica* Andrews
    - subsp. *daurica* (Figs 21.1–21.5)
    - subsp. *coriifolia* (Rupr.) D. Y. Hong (Figs 21.6–21.7)
    - subsp. *mlokosewitschii* (Lomakin) D. Y. Hong (Figs 21.8–21.10)
    - subsp. *macrophylla* (Albov) D. Y. Hong (Figs 21.11–21.15)
    - subsp. *tomentosa* (Lomakin) D. Y. Hong (Figs 21.16–21.18)
    - subsp. *wittmanniana* (Hartwiss ex Lindl.) D. Y. Hong (Figs 21.19–21.21)
22. *Paeonia mascula* (L.) Mill.
    - subsp. *mascula* (Figs 22.1–22.7)
    - subsp. *bodurii* N. Özhatay (Figs 22.8–22.12)
    - subsp. *hellenica* Tzanoud. (Figs 22.13–22.17)
    - subsp. *russoi* (Biv.) Cullen & Heywood (Figs 22.18–22.21)

23. *Paeonia mairei* H. Lév. (Fig. 23.1–23.10)
24. *Paeonia kesrouanensis* (Thiébaut) Thiébaut (Figs 24.1–24.10)
25. *Paeonia coriacea* Boiss. (Figs 25.1–25.2)
26. *Paeonia algeriensis* Chabert (Figs 26.1–26.4)

Subsect. *Paeonia*

Lateral roots fusiform or tuberous. Lower leaves biternate or triternate; leaflets nearly always segmented with leaflets and/or leaf segments (9–)21–340 in number; leaves mostly with bristles along veins above. Flowers solitary and terminal; sepals mostly rounded at the apex.

27. *Paeonia intermedia* C. A. Mey. (Figs 27.1–27.9)
28. *Paeonia tenuifolia* L. (Figs 28.1–28.8)
29. *Paeonia peregrina* Mill. (Figs 29.1–29.7)
30. *Paeonia saueri* D. Y. Hong, X. Q. Wang & D. M. Zhang (Figs 30.1–30.7)
31. *Paeonia arietina* G. Anderson (Figs 31.1–31.11)
32. *Paeonia parnassica* Tzanoud. (Figs 32.1–32.7)
33. *Paeonia officinalis* L.
    subsp. *officinalis* (Figs 33.1–33.5)
    subsp. *banatica* (Rochel) Soó (Figs 33.6–33.8)
    subsp. *huthii* Soldano (Figs 33.9–33.13)
    subsp. *microcarpa* (Boiss. & Reut.) Nym. (Figs 33.14–33.22)

12. ***Paeonia lactiflora*** Pall., *Reise russ. Reich.* 3: 286 (1776).

Paeonia lactiflora Pall. var. *trichocarpa* (Bunge) Stern; *P. albiflora* Pall.

Herbs perennial. Roots thick, cylindrical or carrot-shaped, attenuate toward tip, up to 30 cm long, 2 cm in diameter. Stems up to 1 m tall, usually glabrous. Lower leaves biternate; leaflets and/or leaf segments 10–15, rarely 9, lanceolate or ovate-lanceolate, 4.5–16 cm long, 1.5–6 cm wide, usually with bristles along veins or sometimes glabrous above, glabrous or sparsely pubescent along veins beneath; margins white cartilaginous-thickened, dentate-spinose on the thickenings. Flowers usually 3–4 on a stem, sometimes only terminal one developed with 2–3 axillary sterile buds, single (wild) or double (cultivated), 8–13 cm across. Involucrate bracts 4 or 5, leaf-like. Sepals 3 or 4, broadly ovate or suborbicular, 1–2 cm long, 1–1.7 cm wide, all caudate at the apex. Petals 9–13, white or pink (wild), or various in colour (cultivated), obovate, 3.5–6 cm long, 1.5–4.5 cm wide. Filaments yellow; anthers yellow. Disk yellow or red, 1–5 mm high, waved or incised. Carpels 2–5, green, red or purple, glabrous or rarely sparsely hispid or tomentose; stigmas sessile, red. Follicles ovoid or oblong-ellipsoidal, 2.5–3 cm long, 1.2–1.5 cm in diameter. Seeds black, ovoid-spherical, ca. 7 mm long, ca. 6 mm in diameter.

Flowering from May to early July; fruiting from late July to September.

Chromosome number: 2n = 10.

Growing in bushes and on grasslands, but also in open woods, at an altitude below 2,300 m but up to 3,400 m in Sichuan.

Distributed in East Asia: China, Korean Peninsula, E Mongolia and Russia (the Far East and SE Siberia).

Fig. 12.1. A population in *Corylopsis–Spiraea* thickets at the edges of *Betula* forest, at ca. 1,600 m, Xiaojingou Valley, Mts Daqingshan, Nei Mongol, 2 July 2004. [Voucher: *D. Y. Hong, K. Y. Pan & R. Cao* H04037 (A, BM, CAS, K, MO, PE).]

Fig. 12.2. A population in open *Corylus* thickets, at 1,300 m, Mt Dahaitou, behind the forest farm, Chicheng County, Hebei, 10 June 2003. [Voucher: *D. Y. Hong, K. Y. Pan & Y. Chen* H03002 (PE).]

Opposite:

Fig. 12.3. An inflorescence with four flowers: carpels red, varying from two to four in number, in the population at 400 m, Xinlitun, Xiajiabao, Qingyuan County, Liaoning, 29 May 1998. [Voucher: *D. Y. Hong et al.* H98032 (A, K, MO, PE).]

Fig. 12.4. Polymorphism in petals colour in the population at 1,300 m, Mt Dahaitou, Chicheng County, Hebei, 26 June 2003: **a**, white; **b**, pinkish-white; **c**, pinkish; **d**, pink. [Voucher: *D. Y. Hong, K. Y. Pan & Y. Chen* H03004 (PE).]

Fig. 12.5. A flower with all the sepals caudate at the apex, a yellow disk and four green carpels, in the population shown in Fig. 12.1.

Fig. 12.6. Polymorphism in carpel colour and number in the population shown in Fig. 12.2: **a**, two, green carpels; **b**, four, red-purple carpels.

Fig. 12.7. A flower with three dark purple carpels and a yellowish dentate disk from the population in Mt Huanggangliang, Hexigten (Jingpeng), Nei Mongol, July 2004. [Voucher: *D. Y. Hong, K. Y. Pan et al.* H04040 (A, BM, K, MO, PE).]

Fig. 12.8. Polymorphism in indumentum of carpels in the population shown in Fig. 12.7: **a**, three glabrous carpels; **b**, two sparsely hairy carpels; **c**, three rather densely hairy carpels.

Fig. 12.9. Carrot-shaped roots in the population shown in Fig. 12.2.

12.3
12.4a
12.4b
12.5
12.7
12.4c
12.4d
12.6a
12.6b
12.9
12.8a
12.8b
12.8c

13. ***Paeonia emodi*** Wall. ex Royle, *Ill. bot. Himal. Mts* 57 (1834).

Perennials. Roots carrot-shaped, up to 2.5 cm in diameter. Stems up to 60 cm tall. Lower leaves biternate, with some or all of the 9 leaflets segmented; leaflets and/or leaf segments 15–27, ovate-lanceolate to lanceolate, 7–14 cm long, 1.5–3.8 cm broad, glabrous or with sparse bristles along veins above, always glabrous beneath. Flowers mostly 2–3 on a stem. Involucrate bracts 3–4, leaf-like. Sepals 3–4, green, ovate-orbicular to orbicular, all caudate at the apex. Petals white, 8–10, obovate, often bilobate, ca. 4 cm long, ca. 3 cm broad. Filaments yellow; anthers yellow. Disk pale pink, waved. Carpels single or 2, green, tomentose with hairs 1–2 mm long or glabrous; styles absent or up to 1 mm long; stigmas pink. Follicles long-ovoid or ellipsoid, 2–3.5 cm long, 1.2–1.5 cm in diameter. Seeds brown-black, oblong, 7–9 mm long, 3.5–6 mm in diameter.
Flowering from May to early June; fruiting from July to August.
Chromosome number: 2n = 10, 20.
Growing in bushes on dry or rocky slopes at altitudes from 1,600 m to 3,000 m.
Distributed in the western Himalayas and NE part of the Hindu Kush: S Tibet, N India, Nepal, N Pakistan and NE Afghanistan.

Fig. 13.1 (a & b). A population in open thickets with herbs at 2,480 m, ripe seeds black, Jiangcun Village, Gyirong County, Tibet, 2 August 2001. Photographs by Dr ZHOU Shi-Liang. [Voucher: *S. L. Zhou* H01031 (A, BM, CAS, K, MO, P, PE).]

Opposite:

Fig. 13.2. An individual from the population shown in Fig. 13.1 that was introduced into the Beijing Botanic Garden, Chinese Academy of Sciences, April 2002.

Fig. 13.3. Upper part of a stem of the individual shown in Fig. 13.2, showing one blooming flower and three undeveloped (sterile) flower buds.

Fig. 13.4. Back view of a flower from the population shown in Fig. 13.1, showing sepals that are all caudate at the apex and white petals.

Fig. 13.5. A flower with two carpels from the population shown in Fig. 13.1.

Fig. 13.6. A flower with glabrous carpels and a yellow disk from the population shown in Fig. 13.1.

Fig. 13.7. Carrot-shaped roots from the population shown in Fig. 13.1, 23 September 2006. [Voucher: *F. S. Yang et al.* H06030 (PE).]

13.2
13.3
13.4
13.6
13.5
13.7

14. ***Paeonia sterniana*** H. R. Fletcher, *J. Roy. Hort. Soc.* 84: 327, fig. 103 (1959).

Paeonia emodi Wall. ex Royle subsp. *sterniana* (H. R. Fletcher) Halda

Perennials, 35–60 cm tall, glabrous throughout. Roots carrot-shaped, tap roots up to 2 cm in diameter, more than 30 cm long. Caudex short, multi-branched, and thus many stems caespitose. Lower leaves biternate; leaflets 9, all segmented; leaf segments 20–30 in number, 4–12 cm long, 1.5–3 cm wide, often lobed; lobes acuminate at the apex. Flowers solitary, terminal, but one or two axillary undeveloped (sterile) buds often present, rarely 2 on a stem. Involucrate bracts 2–4 in number, leaf-like. Sepals mostly 3, rarely 4, nearly orbicular, all or mostly caudate at the apex. Petals white to pale rose, obovate, 2.5–3 cm long, 1.5–2 cm wide. Filaments yellow; anthers yellow. Disk less than 1 mm high, waved, green-yellow. Carpels mostly 2, less frequently 3; styles less than 1 mm long, stigmas red. Follicles ovoid, ca. 3 cm long. Seeds ovoid-oblong, black, lucid, 7–8 mm long, ca. 5 mm in diameter.
Flowering from May to early June; fruiting in August.
Chromosome number: 2n = 10.
Growing in forests or thickets at an altitude of 2,830–3,500 m.
Confined to SE Tibet of China.

Fig. 14.1. A population in a *Quercus aquifolioides* forest at 3,000 m, Deba Village, Sumzum Township, Bomi, SE Tibet, 19 May 2006. [Voucher: *D. Y. Hong, Z. Q. Zhou & A. S. Xu* H06017 (PE).]

Fig. 14.2. Close-up of the population shown in Fig. 14.1.

Opposite:

Fig. 14.3. A flower with white petals and three carpels in the population shown in Fig. 14.1.

Fig. 14.4. Another individual in the population shown in Fig. 14.1.

Fig. 14.5. A flower with white petals and two carpels from the population shown in Fig. 14.1.

Fig. 14.6. An individual with two flowers on a stem from the population shown in Fig. 14.1.

Fig. 14.7. Back view of a flower, showing involucrate bracts, sepals that are all caudate at the apex, and a wilting pink petal, in the population shown in Fig. 14.1.

Fig. 14.8. A flower with pink petals and two carpels in the population shown in Fig. 14.1.

Fig. 14.9. Polymorphism in carpel number and colour in the population shown in Fig. 14.1.

Fig. 14.10. Carrot-shaped roots in the population shown in Fig. 14.1.

14.3
14.4
14.5
14.6
14.7
14.8
14.10
14.9a
14.9b
14.9c

15. ***Paeonia anomala*** L., *Mant. pl.* 2: 247 (1771).

Herbs perennial. Tap roots up to 1 m long, thickened, carrot-shaped, attenuate downwards, up to 2 cm in diameter, lateral roots also carrot-shaped, neither tuberous nor fusiform. Lower leaves biternate; leaflets 9; segments 70–100, 2–13 cm long, 0.8–3.2 cm wide, with bristles along veins above, glabrous beneath. Flowers solitary or 2 to 4 on a stem, often only the terminal one fully developed and blooming. Involucrate bracts 1–3, leaf-like. Sepals 3 to 5, mostly caudate at the apex, rarely 1 or very occasionally 2 non-caudate. Petals rose, pale red or red, but rarely white in subsp. *veitchii*, 6–9, obovate, 4–5.5 cm long, 3–4 cm wide. Disk waved, ca. 1.5 mm high. Carpels mostly 3 to 5, densely tomentose, rarely sparsely hairy or glabrous; stigmas sessile or nearly sessile, red. Follicles columnar, 1.5–2.8 cm long, 1–1.2 cm wide. Seeds ovoid or ovoid-spherical, black, 6–7 mm long, 4.5–5 mm wide.
Flowering from late April to middle July; fruiting in August and September.
Chromosome number: 2n = 10.
Two subspecies:

1a. Flowers solitary, rarely 2, or solitary but with one additional undeveloped axillary bud on a stem (Central Asia to the Kola Peninsula via Siberia) . . . . 15a. subsp. *anomala*
1b. Flowers 2–4, rarely solitary but with 1–2 additional undeveloped axillary buds on a stem, very occasionally solitary (central China) . . . . 15b. subsp. *veitchii*

15a. ***Paeonia anomala*** L. subsp. ***anomala***. (Figs 15.1–15.5)

Paeonia altaica K. M. Dai & T. H. Ying; *P. veitchii* subsp. *altaica* (K. M. Dai & T. H. Ying) Halda; *P. sinjiangensis* K. Y. Pan

Growing usually in deciduous or conifers forests in valleys, less frequently in meadows, below 2,400 m in altitude.
Widely distributed from the Kola Peninsula of Russia to the Altai and Baikal: NE Kazakhstan, Mongolia, Russia and N Xinjiang of China.

Fig. 15.1. Habitat: this peony grows well in a *Populus* forest by a stream at 1,200 m on Mt Halamaryi, the Altai, Xinjiang, 3 June 1993. [Voucher: *D. Y. Hong et al.* population No. 3 (PE).]

Fig. 15.2. Habitat: in a *Populus–Betula–Picea* forest in a ravine at 1,060 m, Xiaodonggou, the Altai, Xinjiang, 4 June 1993. [Voucher: *D. Y. Hong et al.* population No. 5 (PE).]

Fig. 15.3. A ready-to-bloom flower in the population shown in Fig. 15.2, showing all the sepals caudate at the apex.

Fig. 15.4. Back view of a flower in the population shown in Fig. 15.1, showing involucrate bracts and sepals.

Fig. 15.5. Carrot-shaped roots in the population shown in Fig. 15.2. Prof. LI Xue-Yu of Shihezi University in Xinjiang is holding the peony.

15b. ***Paeonia anomala*** L. subsp. ***veitchii*** (Lynch) D. Y. Hong & K. Y. Pan in Hong *et al.*, *Novon* 11: 317 (2001). (Figs 15.6–15.17)

Paeonia veitchii Lynch; *P. beresowskii* Kom.; *P. veitchii* Lynch subsp. *veitchii* var. *woodwardii* (Stapf ex Cox) Halda

Flowering from late April to early June; fruiting in August and September.
Chromosome number: 2n = 10.
Growing in forests, grasses on the edges of forests, or bushes, meadows with shrubs, on limestones, at altitudes from 1,800 to 3,870 m.
Widely distributed in central China.

Fig. 15.6. Habitat: an individual by a rock face in a deciduous broad-leaved forest at 2,600 m, N slope, Miyaluo, Sichuan, 12 May 2006. [Voucher: *D. Y. Hong & Z. Q. Zhou* H06006 (PE).]

Fig. 15.7. An individual with four red flowers on a stem in the population shown in Fig. 15.6.

15.8
15.10a
15.10b
15.9
15.11a
15.11b
15.12
15.13

Opposite:

Fig. 15.8. Back view of a flower in the population shown in Fig. 15.6, showing involucrate bracts, and sepals that are all caudate at the apex.

Fig. 15.9. An individual in the population at the edges of forest, 2,700 m, Adi Village, Barkam Town, Barkam, Sichuan, 13 May 2006. [Voucher: *D. Y. Hong & Z. Q. Zhou* H06009 (PE).]

Fig. 15.10 (a & b). An individual with white flowers in a population at 2,900 m, sparse woods, Suomo Township, Barkam, Sichuan, 12 May 2006. [Voucher: *D. Y. Hong & Z. Q. Zhou* H06007 (PE).]

Fig. 15.11 (a & b). An individual with two rose flowers in a population in sparse woods at 2,800 m, Pingqiaogou, Suomo Township, Barkam, Sichuan, 14 May 2006. [Voucher: *D. Y. Hong & Z. Q. Zhou* H06011 (PE).]

Fig. 15.12. An individual from the same population as the plant shown in Fig. 15.11 with pinkish petals.

Fig. 15.13. A red flower in a population at 3,200 m, Mt Yaoshan, Qiaojia County, Yunnan, 26 May 2004. [Voucher: *D. Y. Hong, K. Y. Pan & H. Yu* H04032 (PE).]

Above:

Fig. 15.14. Fruiting peonies at the edge of *Picea–Betula* forest at 3,100 m, stretching from Miyaluo to Mt Zhegu, Lixian County, Sichuan, 19 August 1995. [Voucher: *D. Y. Hong, Y. B. Luo & Y. H. He* H95034 (A, CAS, K, MO, PE, US).]

Fig. 15.15. Polymorphism in indumentum of carpels in the population shown in Fig. 15.14, **a**, glabrous; **b**, hispidulous.

Fig. 15.16. Red follicles and red sepals in the population at 3,000 m in a conifer forest, Jiuzhaigou Valley, Sichuan, 24 August 2006. [Voucher: *D. Y. Hong* H06019 (PE).]

Fig. 15.17. Carrot-shaped roots from the same population as the flower shown in Fig. 15.13.

16. ***Paeonia obovata*** Maxim., *Prim. Fl. amur.* 29 (1859).

Paeonia japonica (Makino) Miyabe & Takeda; *P. oreogeton* S. Moore; *P. vernalis* Mandl.

Herbs perennial, 30–70 cm tall. Roots thick, carrot-shaped. Caudex usually short, 2–8 cm long. Stems often single, always simple, glabrous. Lower leaves spreading or ascending, biternate; leaflets 9, entire, obovate, cuneate at the base, rounded or acute at the apex, 5–20 cm long, 4–14 cm wide, glabrous above, glabrous to densely hispid beneath. Flowers solitary, terminal. Involucrate bracts 1 or 2, leaf-like, rarely lacking. Sepals 2–4, but mostly 3, mostly rounded at the apex. Petals 4–7, spreading or incurved, white, rose, pink-red, red, purple-red, or rarely white with pinkish base or margins, obovate, 3–5.5 cm long, 1.8–2.8 cm wide. Filaments white, whitish-yellow, or purple below but white above to entirely purple; anthers yellow, orange-red or dark purple. Disk yellow, waved, 1–1.5 mm high. Carpels mostly 2 or 3, always glabrous; styles 2–5 mm long; stigmas red. Follicles gradually recurved, ellipsoid, 2–3 cm long. Seeds black, glossy, ovoid-spherical, 6–7 mm long, 5–6 mm in diameter.
Distributed in East Asia.
Two subspecies:

1a. Leaves mostly glabrous or sparsely (occasionally densely) hispid on lower surface; diploid (tetraploids rarely occur in the Far East of Russia and on Mt Changbai of NE China) . . . . 16a. subsp. *obovata*

1b. Leaves usually densely, very occasionally sparsely, hispid on lower surface; tetraploid (the only diploid found in a mixed population in Lushi County, W Henan) . . . . 16b. subsp. *willmottiae*

16a. ***Paeonia obovata*** Maxim. subsp. ***obovata***. (Figs 16.1–16.16)

Flowering from late April to early June; fruiting from late July to August.
Chromosome number: 2n = 10, rarely 2n = 20.
Growing in deciduous broad-leaved, mixed broad-leaved and needle-leaved, or conifer forests below 2,800 m in altitude.

Fig. 16.1. Habitat: a population in a deciduous broad-leaved forest at 1,200 m on Mt Akaboshi, Uma-gun, Doichou-cho, Ehime, Japan, 13 May 1990. [Voucher: *D. Y. Hong* PB90010 (PE).]

Fig. 16.2. Habitat: the same population as shown in Fig. 16.1.

Fig. 16.3. A flower in the population shown in Fig. 16.1.

Fig. 16.4. A pinkish flower in planted *Chamaecyparis–Cryptomeria* forest, at 700 m, To-Uwa-gun, Uwa-Cho, Ehime, Japan, 21 May 1990. [Voucher: *D. Y. Hong* PB90011 (PE).]

Fig. 16.5. A pink flower in the same population as the flower shown in Fig. 16.4.

Fig. 16.6. A violet flower in the same population as the flower shown in Fig. 16.4.

Fig. 16.7. A peony in sparse woods at 1,350 m, Mt West Tianmu, Zhejiang, 8 May 1998. [Voucher: *D. Y. Hong, K. Y. Pan & L. H. Lou* H98001 (A, CAS, K, MO, PE, US).]

Fig. 16.8. A flower with white petals and orange anthers in the same population as the peony shown in Fig. 16.7.

Fig. 16.9. A red-flowered peony in planted sparse *Larix* forest at 650 m, near Lianshanguan Fort, Benxi, Liaoning, 24 May 1998. [Voucher: *D. Y. Hong et al.* H98019 (A, CAS, K, MO, PE, US).]

Fig. 16.10. A red-flowered and leaf-spreading individual in a glade within a conifer forest at 1,500 m, Mt Changbai, Jilin, 27 May 1998. [Voucher: *D. Y. Hong et al.* H98026 (A, CAS, K, MO, PE, US).]

Fig. 16.11. A red flower in the population shown in Fig. 16.10.

Fig. 16.12. A flower with yellow filaments, a pink disk and three carpels, in the population at 1,200 m, Baishilazi Nature Reserve, Kuandian County, Liaoning, 25 May 1998. [Voucher: *D. Y. Hong et al.* H98023 (A, CAS, K, MO, PE, US).]

Fig. 16.13. Another flower from the same population as that shown in Fig. 16.12, but with a yellow disk and two carpels.

Fig. 16.14. Flowers with either one or two carpels on the same plant in the population at 600 m, Xinlitun, Xiaojiabao Township, Qingyuan County, Liaoning, 29 May 1998. [Voucher: *D. Y. Hong et al.* H98031 (A, CAS, K, MO, PE, US).]

Fig. 16.15. Ripe follicles with black and lucid seeds, and aborted red ovules, in the population at 1,350 m, Potou Forest Farm, Miyun County, Beijing, 4 September 1996. [Voucher: *D. Y. Hong, K. Y. Pan & L. Z. Gao* (PE).]

Fig. 16.16. Caudex and carrot-shaped roots in the population shown in Fig. 16.1.

Fig. 16.17. A white-flowered population in an oak forest at 1,880 m, Shangbaiyun, Mt Taibai, Shaanxi, 8 May 1997. [Voucher: *D. Y. Hong, Y. Z. Ye & Y. X. Feng* H97057 (PE).]

Fig. 16.18. A flower with purple filaments in the population shown in Fig. 16.17.

Fig. 16.19. A flower with yellow filaments in the population shown in Fig. 16.20.

Fig. 16.20. A red-flowered population near the population shown in Fig. 16.17, 8 May 1997. [Voucher: *D. Y. Hong, Y. Z. Ye & Y. X. Feng* H97056 (PE).]

16b. *Paeonia obovata* Maxim. subsp. ***willmottiae*** (Stapf) D. Y. Hong & K. Y. Pan in Hong *et al.*, *Pl. Syst. Evol.* 227: 134 (2001). (Figs 16.17–16.22)
Paeonia willmottiae Stapf; *P. obovata* var. *willmottiae* (Stapf) Stern.

Flowering in May; fruiting in August & September.
Chromosome number: 2n = 20.
Growing in deciduous forests at altitudes from 800 m to 2,800 m.
Confined to the mountain area around the Qinling Range: NE Chongqing, W Henan, W Hubei, S Ningxia, Shaanxi, Shanxi and N Sichuan.

Fig. 16.21. A red-flowered individual in the mixed population at 1,550 m, Dakuaidi, Shiziping, Lushi County, Henan, 16 May 1998. [Voucher: *D. Y. Hong et al.* H98006 (A, CAS, K, MO, PE, US).]

Fig. 16.22. Polymorphism of floral parts in the population shown in Fig. 16.21: **a**, white petals and purple anthers; **b**, petals white but with pinkish periphery, anthers orange; **c**, petals red, anthers yellow and filaments white; **d**, filaments red.

Opposite:

Fig. 17.1. A population at 60–80 m, Pollença, Mallorca, the Baleary Islands, Spain, 14 June 2001. [Voucher: *D. Y. Hong & X. Q. Wang* H01025 (A, BM, CAS, K, MO, P, PE).]

Fig. 17.2. An individual at a stony and sparsely grassy site, at 20–150 m in altitude, Soller, Mallorca, 18 April 2001. Photograph by Dr A. Fridlender. [Voucher: *A. Fridlender* H01002 (A, BM, CAS, K, MO, PE).]

Fig. 17.3. A flower from the same population as the plant shown in Fig. 17.2. Photograph by Dr A. Fridlender.

Fig. 17.4. Another individual with a different petal colour from the same population as the plant shown in Fig. 17.2. Photograph by Dr A. Fridlender.

Fig. 17.5. A flower in the same population as the plant shown in Fig. 17.2, showing a leaf that is purple on the abaxial side, purple filaments, yellow anthers and purple carpels. Photograph by Dr A. Fridlender.

Fig. 17.6. Variability in carpel number and colour in the same population as the plant shown in Fig. 17.2: **a**, four purple carpels; **b**, five green carpels; **c**, seven purple carpels; **d**, 10 green carpels. Photographs by Dr A. Fridlender.

Fig. 17.7. Carrot-shaped roots from the population shown in Fig. 17.1.

17. ***Paeonia cambessedesii*** (Willk.) Willk. in Willkomm & Lange, *Prodr. Fl. Hispan.* 3: 976 (1880).

Perennials, glabrous throughout. Caudex occasionally elongated, forming rhizomes up to 15 cm long. Roots carrot-shaped. Stems mostly purple, less frequently green, 30–50 cm tall. Lower leaves biternate; leaflets 9, rarely 8 or 7, never segmented, ovate, oblong, or ovate-lanceolate, 6–11 cm long, 3–6 cm wide, mostly purple beneath. Flowers terminal; involucrate bracts mostly 2. Sepals usually 3, less frequently 2 or 4, rounded but sometimes one of them acute or caudate at the apex, purple. Petals pink, 5–8. Filaments purple, anthers yellow. Disk waved, 1 mm high. Carpels 3–8, but mostly 4–6; styles 1–2 mm long; stigmas 1–1.5 mm wide. Follicles ellipsoid-columnar, 3–4 cm long, 1–1.8 cm wide. Seeds spherical, black, ca. 5 mm in diameter.
Flowering from October to April; fruiting in June and July.
Chromosome number: 2n = 10.
Growing in calcarious soils with shrubs or grasses below 1,400 m altitude.
Endemic to the Balearic Islands, Spain.

18. ***Paeonia corsica*** Sieber ex Tausch, *Flora* 11: 88 (1828).

Paeonia morisii Cesca, Bernardo & Passalacqua; *P. mascula* (L.) Mill. subsp. *russoi* (Biv.) Cullen & Heywood, pro parte.

Perennials. Roots thickened, carrot-shaped, up to 2 cm in diameter. Stems 35–80 cm tall. Petioles glabrous or villose; lower leaves biternate; leaflets 9, rarely segmented nearly to the base, thus leaflets and/or leaf segments mostly 9, less frequently 10, 11, very rarely up to 20, ovate to elliptic, 4–13 cm long, 2–8 cm wide, always glabrous above, but mostly rather densely, less frequently sparsely villose, rarely glabrous beneath. Flowers terminal. Involucrate bracts 0–3, leaf-like. Sepals 2–8, but mostly 3–5, ovate-orbicular, with 1 or 2 (rarely three) acute, the rest rounded at the apex, rarely one caudate at the apex. Petals 7–8, rose. Disk 1 mm high, waved, tomentose on flowers with tomentose carpels. Carpels 1–8, but mostly 2–5, green, purple or red, hispidulous, with hairs 1.5 mm long, brown-yellow, rarely glabrous, the widest above the middle; styles 1.5–3 mm long; stigmas red, 1–1.5 mm wide. Seeds ovoid-spherical, black, ca. 7 mm long, 5–6 mm in diameter.
Flowering from late April to late May; fruiting from late July to September.
Chromosome number: 2n = 10.
Growing in oak and pine forests, maquis and herbs at 400–1,700 m in altitude.
Confined to Corsica (France), Sardinia (Italy) and SW Greece.

Fig. 18.1. A population in a *Pteridium aquilinum* community at the edges of pine forest, at 930 m, Capo dia Media, Galeria, Corsica, 29 May 2001. [Voucher: *D. Y. Hong, X. Q. Wang & A. Fridlender* H01014 (A, CAS, K, MO, PE, UPA).]

Fig. 18.2. A population among low *Santolina* shrubs at 1,230 m, above Oliena, Nuoro, Sardinia, 2 June 2001. [Voucher: *D. Y. Hong, X. Q. Wang & A. Fridlender* H01016 (A, CAS, K, MO, PE).]

Fig. 18.3. An individual in an oak forest at 950–1,000 m, Mt Tonneri, Gennargentu Range, Nuoro, Sardinia, 3 June 2001. The leaves and carpels are clearly shown. [Voucher: *D. Y. Hong, X. Q. Wang & A. Fridlender* H01018 (A, BM, CAS, K, MO, PE, UPA).]

Fig. 18.4. A flowering individual from the same population as the plant in Fig. 18.3.

Fig. 18.5. A pink flower in a sparse pine forest at 850–900 m, Mt Cagna, Corsica, 31 May 2001. [Voucher: *D. Y. Hong, X. Q. Wang & A. Fridlender* H01015 (A, CAS, K, MO, PE, UPA).]

Fig. 18.6. The same flower as shown in Fig. 18.5, showing five glabrous carpels.

Fig. 18.7. Another flower from the same population as that shown in Fig. 18.5, but with three tomentose carpels.

Fig. 18.8. An individual from the population shown in Fig. 18.1 whose flowers range in carpel number continuously from two to five.

Fig. 18.9. Another individual from the population shown in Fig. 18.1 has eight carpels.

Fig. 18.10. An individual with four carpels in a sparse oak forest at 450 m, Skourtou, Akarnaria, Etolia, Greece, 26 May 2002. [Voucher: *D. Y. Hong et al.* H02225 (A, K, MO, PE).]

Fig. 18.11. Carrot-shaped roots from the population featured in Fig. 18.3.

19. ***Paeonia broteri*** Boiss. & Reut., *Diagn. pl. nov. Hisp.* 4. (1842), nom. cons.

Paeonia lusitanica Mill. (1768), nom. rej.

Perennials. Tap roots carrot-shaped, up to 3 cm in diameter, lateral roots thin and basipetally attenuate. Stems glabrous, 30–80 cm tall. Lower leaves biternate, usually most leaflets segmented; leaflets and/or leaf segments 11–32, but mostly 15–21, elliptic or ovate-lanceolate, rarely obovate, 4–10(15) cm long, 1.5–5(6.5) cm wide, usually glabrous on both sides. Flowers terminal. Involucrate bracts 1 or 2, leaf-like. Sepals usually 3, rarely 4, mostly rounded at the apex, up to 3 cm long, 2.6 cm wide, usually glabrous. Petals 6–7, pink-red, 5–6 cm long, 3–4 cm wide. Filaments yellow or purple. Disk waved, 2 mm high, glabrous or tomentose. Carpels mostly 2 or 3, less frequently 1 or 4, tomentose, hairs 2 mm long, rust-brown; stigmas sessile, red, 2.5 mm wide. Follicles 2.5–4 cm long, 1.3–1.6 cm in diameter. Seeds oblong, black, 7–8 mm long, 5–6 mm wide.
Flowering from April to early June; fruiting in August and September.
Chromosome number: 2n = 10.
Growing in bushes, or in oak or pine forests, in limestone soils at altitudes from 300 to 1,830 m.
Confined to the Iberian Peninsula.

Fig. 19.1. A population in a sparse oak forest, Mingorria, Avila, Spain, 10 August 2003. [Voucher: *D. Y. Hong & P. Vargos* H03015 (A, BM, CAS, K, MO, P, PE).]

Fig. 19.2. A flower with two follicles in the population shown in Fig. 19.1.

Opposite:

Fig. 19.3. Three follicles with blue-black seeds from the population shown in Fig. 19.1.

Fig. 19.4. An individual with a rose flower, from the population shown in Fig. 19.1, that was introduced into the Beijing Botanic Garden, Chinese Academy of Sciences, 6 May 2004.

Fig. 19.5. The same flower as that shown in Fig. 19.4.

Fig. 19.6. An individual from a *Pinus sylvestris* forest at 1,200 m, Hoyocasero, Avila, Spain, that was introduced into the Beijing Botanic Garden, Chinese Academy of Sciences, 16 April 2004. [Voucher: *D. Y. Hong & P. Vargos* H03017 (PE).]

Fig. 19.7. The same flower as that shown in Fig. 19.6, showing villous petioles, the abaxial side of sepals, and leaves.

Fig. 19.8. Two slightly different flowers: **a**, from the population shown in Fig. 19.1, showing a yellow disk, yellowish-white filaments and three tomentose carpels; **b**, from the same population as the individual in Fig. 19.6, showing two carpels and filaments that are purple at the lower part, but yellowish-white above.

Fig. 19.9. Carrot-shaped roots from the population shown in Fig. 19.1.

19.4
19.5
19.6
19.7
19.3
19.8a
19.8b
19.9

20. ***Paeonia clusii*** Stern, *Bot. Mag.* 162: t. 9594 (1940).

Perennials totally glabrous except carpels and occasionally lower surface of leaves. Tap roots thickened, slightly fusiform, up to 2.5 cm in diameter, lateral roots carrot-shaped. Lower leaves biternate, all leaflets segmented; segments 23–95, linear, lanceolate to ovate, 5–10 cm long, 0.5–4.5 cm wide. Flowers solitary. Involucrate bracts 1–2, leaf-like. Sepals 3 or 4, ovate-orbicular, mostly rounded at the apex, purple at the periphery. Petals 7, white, 4–5 cm long, 2.5–4 cm wide. Disk flat, 0.5–1 mm high, tomentose. Carpels 2–4, rarely 1, tomentose, hairs 2.0–2.5 mm long, pink or brown-yellow; styles 1–2 mm long; stigmas red, 1.5–2 mm wide. Follicles ellipsoid, ca. 3.8 cm long, 1.5 cm in diameter. Seeds ovoid-spherical, black, ca. 8 mm long, 5 mm in diameter.
Confined to three Aegean islands.
Two subspecies.

1a. Leaf segments 23–95 in number, linear to lanceolate, up to 2.6 cm, very occasionally 3.2 cm wide . . . . 20a. subsp. *clusii*
1b. Leaf segments 23–48 in number, lanceolate to ovate, 2.5–4.5 cm wide . . . . 20b. subsp. *rhodia*

20a. ***Paeonia clusii*** Stern subsp. ***clusii***. (Figs 20.1–20.5)

Flowering from late March to May; fruiting in August.
Chromosome number: 2n = 10, 20.
Adapted to rather dry maquis in limestone areas at altitudes of 200–1,900 m.
Confined to Crete and Karpathos.

20b. ***Paeonia clusii*** Stern subsp. ***rhodia*** (Stearn) Tzanoud., *Cytotax. Stud. Gen. Paeonia* L. *in Greece*, 25 (1977). (Figs 20.6–20.8)

Paeonia rhodia Stearn.

Flowering from late March to May; fruiting in August.
Chromosome number: 2n = 10.
Growing in pine woods at an altitude of 350–850 m.
Confined to Rhodes.

Opposite:

Fig. 20.1 An individual among sparse shrubs with herbs in Crete, Greece, June 1993. Photograph by Dr T. Sang.

Fig. 20.2 An individual in maquis, Crete, Greece, June 1993. Photograph by Dr T. Sang.

Fig. 20.3 An individual among low sparse shrubs in Crete, Greece, May 1980.

Fig. 20.4 The same plant as that shown in Fig. 20.3.

Fig. 20.5 Thickened tap root in a population from Crete, 23 April 2003. Photograph by Mr N. Turland.

Fig. 20.6 An individual with several fruiting flowers, carpels from two to four in number, from Rhodes, Greece, June 1993. Photograph by Dr T. Sang.

Fig. 20.7 An individual with five carpels in sparse woods, Rhodes, Greece, June 1993. Photograph by Dr T. Sang.

Fig. 20.8 An individual with five carpels, also Rhodes, Greece, June 1993. Photograph by Dr T. Sang.

20.1
20.2
20.3
20.4
20.5
20.6
20.7
20.8

21. ***Paeonia daurica*** Andrews, *Bot. Repos.* 7: t. 486 (1807).

Paeonia triternata Pall. ex DC.; *P. mascula* (L.) Mill. subsp. *triternata* (Pall. ex DC.) Stearn & P. H. Davis.

Perennials. Roots carrot-shaped, up to 4.6 cm in diameter. Lower leaves biternate; leaflets 9, occasionally one or two segmented, and thus leaflets plus leaf segments usually 9, rarely 10 or 11, entire, broad-obovate, oblong, rarely wide-elliptic, sometimes undulate, truncate or rounded, rarely acute or even short-acuminate at the apex, 8–17 cm long, 4.8–11.5 cm wide, glabrous above, glabrous or sparsely to densely villose, or sparsely to rather densely puberulous beneath. Flowers terminal. Involucrate bracts 0–2, leaf-like. Sepals 2–3, all rounded at the apex. Petals 5–8, usually red or rose, less yellow, pale yellow, white, or yellow but with a red spot at the base or with red periphery. Disk ca. 1 mm high, waved, tomentose above. Carpels 1–5 but mostly 2 or 3, mostly tomentose, less frequently glabrous, hairs 2.5–3 mm long; stigmas nearly sessile, red, 1.5–2 mm wide.
Distributed widely from NE Iran to Croatia via the Caucasus and Turkey.
Seven subspecies, of which six are represented here.

1a. Sepals often villose on the abaxial side; leaves rather densely villose on the lower side . . . . 21f. subsp. *velebitensis* D. Y. Hong
1b. Sepals glabrous; leaves sparsely villose or puberulous, less frequently densely villose on the lower side or glabrous
  2a. Carpels glabrous or nearly glabrous; petals yellow
    3a. Leaves densely villose and thus greyish beneath . . . . 21c. subsp. *macrophylla* (Albov) D. Y. Hong
    3b. Leaves usually sparsely villose beneath . . . . 21g. subsp. *wittmanniana* (Hartwiss ex Lindl.) D. Y. Hong
  2b. Carpels tomentose; petals red, rose, white or yellow
    4a. Leaflets or leaf segments puberulous or glabrous beneath, obovate, apex rounded or obtuse, often with a short mucro . . . . . . 21d. subsp. *mlokosewitschii* (Lomakin) D. Y. Hong
    4b. Leaflets or leaf segments villose or glabrous beneath, obovate, oblong or wide-elliptic, apex rounded to short-acuminate
      5a. Petals red or rose; leaves glabrous or sparsely villose beneath
        6a. Leaflets or leaf segments broad-obovate, truncate to rounded at apex. . . . . 21a. subsp. *daurica*
        6b. Leaflets or leaf segments obovate to oblong, rounded to acute at apex . . . . 21b. subsp. *coriifolia* (Rupr.) D. Y. Hong
      5b. Petals yellow or yellowish white, but sometimes red at periphery or with a red spot at base; leaves villose beneath
        7a. Leaves mostly densely villose and thus greyish beneath . . . . 21e. subsp. *tomentosa* (Lomakin) D. Y. Hong
        7b. Leaves usually sparsely villose beneath. . . . . 21g. subsp. *wittmanniana* (Hartwiss ex Lindl.) D. Y. Hong

21a. ***Paeonia daurica*** Andrews subsp. ***daurica***. (Figs 21.1–21.5)

Flowering from middle April to early June; fruiting in August and September.
Chromosome number: 2n = 10.
Growing in woods at altitudes of 350–1,550 m.
Distribution ranges from Turkey to Crimea and Croatia.

Fig. 21.1 A rose-flowered peony growing in an oak forest at 880–920 m, Havza, Samsun, Turkey, 21 May 2002. [Voucher: *D. Y. Hong et al.* H02221 (A, CAS, K, MO, PE, UPA).]

Fig. 21.2 A red-flowered peony in the same population as that shown in Fig. 21.1, showing lower leaves with 9 obovate leaflets.

Fig. 21.3 The same flower as that shown in Fig. 21.2.

Fig. 21.4 A flower in a *Fagus orientalis* forest at 1,290 m, Mt Amanos, Hatay, Turkey, showing sepals, a yellow and dentate disk, and two lanate carpels, 18 May 2002. [Voucher: *D. Y. Hong et al.* H02213 (MO, PE).]

Fig. 21.5 Carrot-shaped roots from the same population as the flower in Fig. 21.4.

Fig. 21.6 A purplish-flowered population in *Quercus–Cornus–Acer* forest, 630 m, Igoeti, Kartli, Georgia, 2 May 1999. [Voucher: *D. Y. Hong & S. L. Zhou* H99029 (A, CAS, K, MO, PE, US).]

Fig. 21.7 A red flower from the population shown in Fig. 21.6.

21b. *Paeonia daurica* Andrews subsp. ***coriifolia*** (Rupr.) D. Y. Hong, *J. Linn. Soc. Bot.* 143: 145 (2003). (Figs 21.6–21.7)
Paeonia caucasica (Schipcz.) Schipcz.; *P. ruprechtiana* Kem.-Nath.

Flowering from late April to May; fruiting in August.
Chromosome number: 2n = 10.
Growing in deciduous or mixed deciduous and conifer forests, below 1,100 m in altitude.
Confined to the W and NW Caucasus.

21c. *Paeonia daurica* Andrews subsp. ***macrophylla*** (Albov) D. Y. Hong, *J. Linn. Soc. Bot.* 143: 147 (2003). (Figs 21.8–21.12)
Paeonia macrophylla (Albov) Lomakin; *P. steveniana* Kem.-Nath.; *P. wittmanniana* Steven, pro parte.

Flowering from late May to June; fruiting in August and September.
Chromosome number: 2n = 20.
Growing in deciduous forests, or mixed deciduous and conifer forests, at altitudes from (800)1,160 m to 2,400 m.
Distributed in the mountainous area in Armenia, SW Georgia and NE Turkey.

21.8 21.9 21.10 21.11 21.12

21d. *Paeonia daurica* Andrews subsp. ***mlokosewitschii*** (Lomakin) D. Y. Hong, *J. Linn. Soc. Bot.* 143: 146 (2003). (Figs 21.13–21.15)

Paeonia mlokosewitschii Lomakin; *P. lagodechiana* Kem.-Nath.

Flowering from April to May; fruiting in August and September.
Chromosome number: 2n = 10.
Growing in deciduous forests at altitudes of 960–1,060 m.
Confined to a small area in E Georgia, NW Azerbaijan and Dagestan of Russia.

21e. *Paeonia daurica* Andrews subsp. ***tomentosa*** (Lomakin) D. Y. Hong, *J. Linn. Soc. Bot.* 143: 148 (2003). (Figs 21.16–21.18)

Paeonia tomentosa (Lomakin) N. Busch; *P. wittmanniana* Hartwiss ex Lindl. subsp. *tomentosa* (Lomakin) N. Busch.

Flowering in May and early June; fruiting in August.
Chromosome number: 2n = 20.
Growing in deciduous forests at altitudes from 1,170 to 2,740 m.
Its area ranges from Azerbaijan in the Talish Mountains to NE Iran in the Elburz Mountains.

21f. *Paeonia duarica* subsp. ***velebitensis*** D. Y. Hong, Peonies of the World Vol. 1: 178 (2010).
Flowering unknown; fruiting unknown.
Chromosome number: unknown.
Growing at altitudes of 900–1,150 m.
Confined to the Velebit Mts, Croatia.

21g. *Paeonia daurica* Andrews subsp. ***wittmanniana*** (Hartwiss ex Lindl.) D. Y. Hong, *J. Linn. Soc. Bot.* 143: 146 (2003). (Figs 21.19–21.21)

Paeonia wittmanniana Hartwiss ex Lindl.; *P. abchasia* Miscz. ex Grossh.

Flowering from late April to May; fruiting in August and September.
Chromosome number: 2n = 20.
Growing in deciduous forests and alpine or sub-alpine meadows at altitudes of (800)1,500–2,300 m.
Confined to NW Georgia and the adjacent region of Russia.

Opposite:

Fig. 21.8 A flower from a population with flowers all yellowish in a *Fagus orientalis* forest at 1,160 m, Machuntseti, Adjaria, Georgia, 21 May 1999. [Voucher: *D. Y. Hong & S. L. Zhou* H99065 (A, CAS, K, MO, PE, US).]

Fig. 21.9 A flower from the population featured in Fig. 21.8.

Fig. 21.10 Another flower with green sepals, a yellow disk, purple filaments, and two glabrous carpels from the population featured in Fig. 21.8.

Fig. 21.11 A yellow-flowered individual in a *Fagus orientalis* forest at 1,700 m, Bakuriani, Kartli, Georgia, June 1999. [Voucher: *D. Y. Hong & S. L. Zhou* H99060 (A, CAS, K, MO, PE, US).]

Fig. 21.12 Another flower with yellow petals that have a pinkish periphery from the population shown in Fig. 21.14.

Fig. 21.13 A population in a deciduous broad-leaved forest at 1,040 m, Lagodekhi, Kachetia, Georgia, 3 May 1999 [Voucher: *D. Y. Hong & S. L. Zhou* H99035 (A, CAS, K, MO, PE, US)]: **a**, petals yellowish-white; **b**, petals yellowish-white but with a pinkish band; **c**, petals white but with a pink periphery; **d**, petals red.

Fig. 21.14 Individuals with differently coloured petals grow together in the population shown in Fig. 21.13.

Fig. 21.15 Typical polymorphism in flower colour in the population shown in Fig. 21.13 .

Fig. 21.16 A flower from a population with yellowish flowers in sparse and low *Fagus–Quercus* woods at 1,170 m, Orand, Lerik, Azerbaijan, 9 May 1999. [Voucher: *D. Y. Hong & S. L. Zhou* H99046 (A, CAS, K, MO, PE, US).]

Fig. 21.17 A flower with a yellow disk and three lanate carpels from the population featured in Fig. 21.16.

Fig. 21.18 Carrot-shaped roots from the population featured in Fig. 21.16.

Fig. 21.19. An individual in the Tbilisi Botanic Garden, introduced from Abchasia, showing petals white but with a pinkish band, 1 May 1999. [Voucher: *D. Y. Hong & S. L. Zhou* H99025 (PE).]

Fig. 21.20. A plant in the Bakuriani (Georgia) Botanic Garden, introduced from Abchasia, showing white petals, 19 May 1999. [Voucher: *D. Y. Hong & S. L. Zhou* H99063 (PE).]

Fig. 21.21. An individual in the Tbilisi Botanic Garden, introduced from Abchasia, showing a yellow and dentate disk and two tomentose carpels, 23 May 1999. [Voucher: *D. Y. Hong & S. L. Zhou* H99068 (PE).]

22. ***Paeonia mascula*** (L.) Mill., *Gard. dict. ed. 8.* No.1 (1768).

Herbs perennial up to 80 cm tall. Roots always carrot-shaped. Stems always glabrous, up to 1 cm in diameter. Lower leaves biternate with some leaflets segmented; leaflets plus leaf segments usually 11–15, up to 21, ovate, obovate, wide-elliptic or oblong, acute at the apex, 4.5–18 cm long, 3–9 cm wide, glabrous above, mostly glabrous, sometimes hispid beneath. Flowers terminal. Involucrate bracts 1 or 2, leaf-like. Sepals 3 or 4, all rounded or only one acute at the apex, purple or purple at the periphery, glabrous. Petals 7–9, pink, red, white or white with pink shade at the base or periphery. Disk waved, 1 mm high, tomentose; carpels mostly 3 or 4, rarely 2, 1 or 5, always tomentose, hairs 3 mm long; styles absent or 1 mm long; stigmas 2–2.5 mm wide, red. Follicles ovoid-columnar. Chromosome number: 2n = 20.
Widely distributed from N Spain to N Iraq via France, Italy, the Balkans, Cyprus and Turkey.
Four subspecies.

1a. Flowers red or pink (entire range, except some Aegean islands, Çanakkale Province (NW Turkey), Sicily (Italy) and Calabria (Italy)) . . . . 22a. subsp. *mascula*
1b. Flowers mostly white, rarely red or pink
  2a. Leaves nearly always hispid, very rarely glabrous beneath (Sicily and Calabria) . . . . 22d. subsp. *russoi*
  2b. Leaves mostly glabrous, rarely very sparsely hispid beneath
    3a. Leaflets plus leaf segments of lower leaves 9–11 in number, 13–17.5 × 7–9 cm (Çanakkale) . . . . 22b. subsp. *bodurii*
    3b. Leaflets plus leaf segments of lower leaves (9–)11–14(–21) in number, 9–13 × 4.5–7.5 cm (S Greece and Aegean islands). . . . . .22c. subsp. *hellenica*

22a. ***Paeonia mascula*** (L.) Mill. subsp. ***mascula***. (Figs 22.1–22.7)

Flowering from April to May; fruiting in August and September.
Growing in deciduous forests or in thickets at altitudes from 300 to 2,200 m.
Covering nearly the entire range of the species.

22.1 22.2

Opposite:

Fig. 22.1 A population in a sparse *Quercus–Acer* wood at 360 m, Monton, Chadieu, Clermont-Ferrand, France, 4 April 2001. Photograph by Dr A. Fridlender. [Voucher: *A. Fridlender* H01004a (PE).]

Fig. 22.2 An individual in the population shown in Fig. 22.1, 30 April 2001. Photograph by Dr A. Fridlender.

Above:

Fig. 22.3 A peony with red flowers in Cole d'Or, France, 28 April 2001. Photograph by Dr A. Fridlender. [Voucher: *A. Fridlender* H01003 (PE).]

Fig. 22.4 Another flower from the same population as that in Fig. 22.3, showing purple-red filaments, a lanate disk and four lanate carpels. Photograph by Dr A. Fridlender.

Fig. 22.5 A population in an oak forest at 1,130 m, Serra la Neviera, Potenza, Italy, 7 June 2001. [Voucher: *D. Y. Hong, X. Q. Wang & A. Fridlender* H01023 (A, CAS, K, MO, PE, UPA).]

Fig. 22.6 An individual showing two lanate carpels from the population shown in Fig. 22.5.

Fig. 22.7 Carrot-shaped roots from the population shown in Fig. 22.5.

22b. *Paeonia mascula* (L.) Mill. subsp. ***bodurii*** N. Özhatay in Özhatay & Özhatay, *Karaca Arb. Mag*. 3 (1): 21, figs 3–5 (1995). (Figs 22.8–22.12)

Flowering from late April to May.
Growing in sparse secondary forests.
Confined to a small area near Çamyayla in Çanakkale Province, NW Turkey.

22c. *Paeonia mascula* (L.) Mill. subsp. ***hellenica*** Tzanoud., *Cytotax. Stud. Gen. Paeonia* L. *in Greece*, 36 (1977). (Figs 22.13–22.17)
Paeonia mascula subsp. *icarica* Tzanoud.

Flowering from April to middle May; fruiting in August.
Growing mostly in forests or in sparse *Crataegus* thickets on limestones at an altitude of 500–2,000 m.
Confined to S Greece and the Aegean islands.

Fig. 22.8 A population in a sparse secondary forest at 620 m, Çamyayla, Çanakkale, Turkey, 12 May 2002. [Voucher: *D. Y. Hong et al.* H02203 (A, CAS, K, MO, PE, UPA).]

Fig. 22.9 An individual with white flowers in the population shown in Fig. 22.8.

Fig. 22.10 Another plant from the population shown in Fig. 22.8 with petals that are white but pinkish at the base, purple-red filaments and three lanate carpels.

Fig. 22.11 A back view of the flower shown in Fig. 22.10, showing one involucrate bract and four sepals.

Fig. 22.12 Carrot-shaped roots from the population shown in Fig. 22.8.

Fig. 22.13 A population with abundant individuals in a *Pteridium aquilinum* community with scattered trees at 880 m, Mt Ochi, Euboea Island, Greece, 28 May 2002. [Voucher: *D. Y. Hong, D. M. Zhang & X. Q. Wang* H02226 (A, BM, CAS, K, MO, P, PE).]

Fig. 22.14 Polymorphism in carpel number in the population shown in Fig. 22.13: **a**, two carpels; **b**, three carpels; **c**, four carpels.

Fig. 22.15 An individual from the population shown in Fig. 22.13 but introduced into Beijing Botanic Garden, Chinese Academy of Sciences, showing petals that are white but rose at the base, April 2003.

Fig. 22.16 Carrot-shaped roots from the population shown in Fig. 22.13.

Fig. 22.17 The individual shown in Fig. 22.15, showing dark blue seeds and aborted red ovules, August 2003.

22d. *Paeonia mascula* (L.) Mill. subsp. ***russoi*** (Biv.) Cullen & Heywood, *Feddes Repert.* 69: 35 (1964). (Figs 22.18–22.21)

Flowering from April to middle May; fruiting in August.
Growing in sparse deciduous forests or *Crataegus* thickets on limestones at altitudes from 800 to 1,650 m.
Confined to Sicily and Calabria, Italy.

Fig. 22.18 A small population among exposed calcareous rocks at 1,240 m, Mt Pizzuta, Sicily, 4 June 2001. [Voucher: *D. Y. Hong & X. Q. Wang* H01019 (PE).]

Fig. 22.19 A rather large population among calcareous rocks at the edges of *Fagus sylvestris* forest at 1,630 m, Mt Carbonara, Madonie Mts, Sicily, 5 June 2001. [Voucher: *D. Y. Hong, X. Q. Wang & A. Fridlender* H01020 (A, CAS, K, MO, PE, UPA).]

Fig. 22.20 An individual in the population shown in Fig. 22.19, with three carpels, a purple-red stem and petioles.

Fig. 22.21 Carrot-shaped roots from an individual in the population shown in Fig. 22.19.

23. *Paeonia mairei* H. Lév., *Bull. Acad. Int. Géogr. Bot.* 25: 42 (1915).

Perennials up to 1 m tall. Roots carrot-shaped. Caudex (rhizomes) up to 15 cm long. Stems single, simple, glabrous. Lower leaves biternate; leaflets 9, some segmented; leaflets plus leaf segments 13–24, mostly 14–17, oblong-ovate or oblong-lanceolate, 6–16.5 cm long, 1.8–7 cm wide, glabrous, usually acuminate or even caudate at the apex. Flowers terminal. Involucrate bracts 1–3, leaf-like or linear. Sepals 3–5, all rounded or sometimes one caudate at the apex. Petals 7–9, pink to red, obovate, 3.5–7 cm long, 2–4.5 cm wide. Disk yellow, annular, ca. 1 mm high. Carpels 2 or 3, rarely 1, sparsely to densely yellow papillate to hispidulous, sometimes glabrous; styles up to 4 mm long; stigmas red, 1.2–1.5 mm wide. Follicles ellipsoid, 2–3 cm long, 1–1.2 cm in diameter. Seeds black, oblong-spherical, 7–8 mm long, 4–5 mm in diameter.

Flowering in April and May; fruiting from August to September.

Chromosome number: 2n = 20.

Growing in deciduous broad-leaved forests on limestones at an altitude of (800)1,200–3,400 m.

Endemic to China: Chongqing, SE Gansu, S Shaanxi, C & S Sichuan and NE Yunnan.

Fig. 23.1 A population in a mixed deciduous broad-leaved and conifer forest at 2,000 m, the Qingling Range, between Ningshaan and Xi'an, Shaanxi, 7 May 1997. [Voucher: *D. Y. Hong, Y. Z. Ye & Y. X. Feng* H97053 (A, CAS, K, MO, PE,US).]

Fig. 23.2 An individual in the population shown in Fig. 23.1.

Fig. 23.3 A pale rose flower in the population shown in Fig. 23.1.

Fig. 23.4 The flower shown in Fig. 23.3, back view showing one involucrate bract and two sepals (one of them actually intermediate between bract and sepal).

Fig. 23.5 An individual with a pale purple flower in a secondary *Larix–Juglans* forest at 1,860 m, the Wolong Nature Reserve, Sichuan, 9 May 1985. [Voucher: *D. Y. Hong & X. Y. Zhu* PB85023 (PE).]

Fig. 23.6 A rose-flowered individual in thickets at 2,500 m, Qiaoqi, Baoxing County, Sichuan, 20 May 1982. [Voucher: *D. Y. Hong & Z. H. Zhong* PB82122 (PE).]

Fig. 23.7 The unique hispidulous hairs on carpels, green disk, and polymorphism of carpel number in a deciduous broad-leaved forest at 1,300 m, Honghe Valley, Mt Taibai, Shaanxi, 22 June 2004: **a**, a single carpel; **b**, three carpels; **c**, four carpels. [Voucher: *D. Y. Hong & Y. Ren* H04036 (PE).]

Fig. 23.8 Two flowers in a secondary forest at 2,200 m, N slope, Tudiliang, Maoxian, Sichuan, 11 May 2006. [Voucher: *D. Y. Hong* & *Z. Q. Zhou* H06005 (PE).]

Fig. 23.9 Caudex and roots: **a**, a young individual in Tangdan, Dongchuan, Yunnan, with a short caudex and one carrot-shaped root, 24 May 2004. [Voucher: *D. Y. Hong & H. Yu* H04031 (PE).]; **b**, an individual in the same population as the carpels shown in Fig. 23.7, showing a long caudex and carrot-shaped roots.

Fig. 23.10 Lower part of an individual from the same population as that featured in Fig. 23.8, showing caudex, roots and a single stem.

24. ***Paeonia kesrouanensis*** (Thiébaut) Thiébaut, *Fl. Lib.-Syr.* (in *Mém. Inst. Égypte, Cairo*, 31) 37 (1936).
Paeonia turcica P. H. Davis & Cullen

Perennials. Roots carrot-shaped. Stems 35–80 cm tall, glabrous. Lower leaves biternate; leaflets and/or leaf segments 10–14, rarely up to 17, ovate or elliptic, acute at the apex, 8–18 cm long, 4–7.5 cm wide, glabrous above, sparsely to rather densely villose beneath. Flowers solitary. Involucrate bracts 0–2 but mostly one. Sepals 3–4, less frequently 2 or 5, mostly or all rounded but sometimes one acute at the apex, usually glabrous, rarely sparsely hirsute outside. Petals 5–9, pale pink, pink or red. Disk 1 mm high, entire, rarely waved. Carpels 1–3, rarely 4 or 5, glabrous; styles 1.5–3.5 mm long, straight part of styles and stigmas 3.5–7.0 mm long.
Flowering from late April to late May; fruiting from late July to September.
Chromosome number: 2n = 20.
Growing in forests or in *Quercus* scrubs at altitudes of (800)1,000–1,800 m.
Distributed in Lebanon, SW Syria, and SW and S Turkey.

Fig. 24.1 A large population in a *Pinus nigra* forest at 1,450 m, Mt Boz Dağ, Acipayam, Denizli, Turkey, 15 May 2002. [Voucher: *D. Y. Hong et al.* H02207 (A, CAS, K, MO, PE, UPA).]

Fig. 24.2 An individual in the population shown in Fig. 24.1.

Fig. 24.3 A large population with abundant individuals in a *Cedrus libani* forest at 1,570 m, above Patlangac, S of Elmali, Antalya, Turkey, 15 May 2002. [Voucher: *D. Y. Hong et al.* H02208 (A, BM, CAS, K, MO, PE).]

Fig. 24.4 A flower of the individual shown in Fig. 24.2.

Fig. 24.5 Another flower from the individual shown in Fig. 24.2, back view, showing one involucrate bract and four sepals.

Opposite:

Fig. 24.6 Polymorphism in carpel number and colour in the population shown in Fig. 24.1: **a**, two, greenish carpels; **b**, three, dark-purple carpels; **c**, four, pure green carpels; **d**, five, dark-purple carpels.

Fig. 24.7 An individual in the population shown in Fig. 24.2 showing flowers with either two or four carpels.

Fig. 24.8 A monocarpellary flower with a yellow disk in the population at 1,180 m, above Senkoy, Hatay, Turkey, 17 May 2002. [Voucher: *D. Y. Hong et al.* H02210 (A, BM, CAS, K, MO, PE, UPA).]

Fig. 24.9 An individual in a *Fagus orientalis* forest at 1,380–1,540 m, Mt Amanos, Hatay, Turkey, showing a red disk and two carpels, 18 May 2002. [Voucher: *D. Y. Hong et al.* H02214 (A, BM, CAS, K, MO, PE, UPA).]

Fig. 24.10 Carrot-shaped roots from the same population as the flower in Fig. 24.8.

24.6a
24.6b
24.6c
24.6d
24.7
24.8
24.9
24.10

25. ***Paeonia coriacea*** Boiss., *Elench. Pl. nov.* 7 (1838).

Paeonia mascula (L.) Mill. subsp. *coriacea* (Boiss.) Malagarriga

Perennials glabrous throughout, very occasionally hairy on leaves and carpels. Roots carrot-shaped. Lower leaves biternate; leaflets 9, with 1 or several segmented; leaflets plus leaf segments 10–15, ovate-orbicular or broad-ovate, acute, less frequently obtuse or short-acuminate at the apex, 5–15 cm long, 2–8 cm wide, glabrous, very occasionally puberulous on the lower surface. Flowers terminal. Involucrate bracts usually 1 or 2, leaf-like. Sepals usually 3, less frequently 2, all or mostly rounded at the apex, purple. Petals red. Disk waved, 1–2 mm high, red. Carpels 1–4, but mostly 2, glabrous, occasionally very sparsely hirsute; styles 1.5–3 mm long; stigmas red, 2–2.5 mm wide. Follicles columnar, 3.5–4.8 cm long. Seeds oblong, black, 7–8 mm long, 5–6 mm in diameter.
Flowering from late April to early June; fruiting in September.
Chromosome number: 2n = 20.
Growing in woods of *Quercus* or *Cedrus* in limestone areas at an altitude of 600–2,100 m.
Distributed in S Spain and Morocco.

Fig. 25.1 An individual in a sparse wood at 1,450 m, Sierra de Alfacar, Granada, Spain, 15 August 2003. [Voucher: *D. Y. Hong & A. Quintanar* H03018 (A, CAS, K, MO, PE).]

Fig. 25.2 Two nearly mature follicles of the individual shown in Fig. 25.1.

Opposite:

Fig. 26.1 A population *D. Y. Hong et al.* H09001 on the edges of *Cedrus atlantica* Manetti forests at Talla-Guild, Boghni, Babor Mountain, Algeria. (Photograph by Dr H. Laouer; thanks to Dr D. Harzallah).

Fig. 26.2 An individual with a pink flower in the population shown in Fig. 26.1. (Photograph by Dr H. Laouer; thanks to Dr D. Harzallah).

Fig. 26.3 Another individual with a red flower in the population shown in Fig. 26.1. (Photograph by Dr H. Laouer; thanks to Dr D. Harzallah).

Fig. 26.4 A flower with a single carpel (the most frequent form) in the population shown in Fig. 26.1. (Photograph by Dr H. Laouer; thanks to Dr D. Harzallah).

26. *Paeonia algeriensis* Chabert, *Bull. Soc. Bot. France* 36: 18 (1889).

Paeonia corallina Retz. var. *atlantica* Coss.; *P. corallina* Retz. subsp. *atlantica* (Coss.) Maire; *P. mascula* (L.) Mill. subsp. *atlantica* (Coss.) Greuter & Burdet; *P. coriacea* Boiss. var. *atlantica* (Coss.) Stern.

Perennials. Roots carrot-shaped. Stems more than 50 cm tall, 0.7–1.0 cm in diameter. Lower leaves biternate with one or several of 9 leaflets segmented and thus leaflets plus leaf segments 10–13; petioles always villose; leaflets and/or leaf segments ovate or oblong, rounded or broad-cuneate at the base, acute at the apex, 9–18 cm long, 5.5–9.5 cm broad, always moderate to densely white-villose beneath. Flowers solitary and terminal. Involucrate bracts 1, leaf-like, or absent; sepals 3 or 4 in number, orbicular or oblong-orbicular, all rounded at the apex, glabrous, purple inside and at the periphery, 2.5–3.5 cm long, 2–2.5 cm wide. Petals pink or red, 5–6 cm long, 3–4 cm wide, rounded at the apex. Disk ca. 1 mm high, slightly waved, glabrous. Carpels mostly single, less frequently 2 in number, nearly always glabrous, very occasionally sparsely hairy; styles 1–3 mm long; stigmas red, 2–3 mm wide. Follicles columnar, 4–5.4 cm long. Seeds black, ovoid-oblong, ca. 9 mm long, 7.5 mm in diameter, seed-coat foveolate.

Flowering from middle May to late June.

Chromosome number: unknown (the only species of *Paeonia* whose chromosome number is still unknown).

Growing in broad-leaved or mixed broad-leaved and coniferous forests with calcarious soils at altitudes of 1,100–2,000 m.

A narrow endemic of Kabylie (Algeria), N Africa; found only along the coastal mountain range, concretely on Mts Babor, Magris, Djurdjura, and in the vicinity of Tala-Kitane and Kefrida.

26.1 26.2 26.3 26.4

27. *Paeonia intermedia* C. A. Mey. in Ledebour, Meyer & Bunge, *Fl. altaic*. 2: 277 (1830).

Paeonia anomala L. subsp. *intermedia* (C. A. Mey.) Trautv.; *P. intermedia* C. A. Mey. subsp. *pamiroalaica* Ovcz.; *P. anomala* L. subsp. *pamiroalaica* (Ovcz.) R. Cooper.

P. anomala L. subsp. *hybrida* Halda, non Pallas.

Herbs perennial, up to 70 cm tall. Tap roots cylindrical, to 2 cm in diameter; lateral roots tuberous, tubers spheroidal to long-fusiform. Lower leaves biternate, with bristles along veins above, always glabrous beneath; leaflets several times segmented; segments 70–100, linear, 6–16 cm long, 0.4–1.8 cm wide, sometimes lobed, acuminate at the apex. Flowers terminal. Involucrate bracts 3, leaf-like. Sepals 3–5, mostly rounded (at least 2 non-caudate) at the apex, glabrous. Petals 7–9, purple-red. Disk annular, incised, up to 2.5 mm high. Carpels 2–5, but mostly 3, tomentose, rarely glabrous; stigmas sessile, 1 mm wide, red. Follicles ellipsoid or ovoid-ellipsoid, 2–2.8 cm long, 1.1–1.3 cm wide. Seeds black, glossy, long-ovoid, 5–5.5 mm long, 3–3.5 mm in diameter.

Flowering from late May to June; fruiting from August to September.

Chromosome number: 2n = 10.

Growing in grassy or shrubby slopes, meadows, steppes, rarely in open woods.

Widely distributed in N Xinjiang of China, Kazakhstan, Kirghizia, Tajikistan, Uzbekistan, and the Altai of Russia.

Fig. 27.1 A large population in open *Berberis–Spiraea* bushes at 1,300 m, Mt Halamaryi, the Altai, Xinjiang, 3 June 1993. [Voucher: *D. Y. Hong et al.* Population No. 2 (PE).]

Opposite:

Fig. 27.2 A population in a sparse *Prunus–Malus* wood at 1,150 m, Jartoulong Valley, Yining, Xinjiang, 31 May 1993. [Voucher: *D. Y. Hong et al.* Population No. 1 (PE).]

Fig. 27.3 An individual from the population shown in Fig. 27.1.

Fig. 27.4 Another large individual from the population shown in Fig. 27.1.

Fig. 27.5 A flower from the population shown in Fig. 27.1.

Fig. 27.6 A flower from a plant growing in *Berberis–Spiraea* bushes, at 1,050 m, Xiaodonggou, the Altai, Xinjiang, back view, showing sepals all rounded at the apex, 4 June 1993. [Voucher: *D. Y. Hong et al.* Population No. 4 (PE).]

Fig. 27.7 An individual from the population shown in Fig. 27.1, showing flowers with two, three or five dark-purple and glabrous carpels, and pink petals.

Fig. 27.8 A flower from the population shown in Fig. 27.2, showing five hairy carpels, and three of five sepals rounded at the apex.

Fig. 27.9 Tuberous roots from a plant in the population shown in Fig. 27.2.

27.2 27.4 27.3 27.5 27.6 27.8 27.7 27.9

28. *Paeonia tenuifolia* L., *Syst. Nat. ed. 10*, 2: 1079 (1759).
Paeonia biebersteiniana Rupr.; *P. carthalinica* Ketsk.; *P. lithophila* Kotov.

Perennials 18–60 cm tall. Tap roots elongated, lateral roots always tuberous, with tubers fusiform, oblong or even spherical. Caudex branched, 2–6 cm long; stems glabrous. Lower leaves triternate; leaflets segmented several times; segments 134–340, linear or filiform, final segments 0.5–3.8 cm long, 0.5–8 mm wide, glabrous on both sides, but sometimes covered with bristles along veins above. Flowers terminal. Involucrate bracts 1–3, leaf-like. Sepals 4 or 5, rarely 3, all rounded or one of them caudate at the apex, 1–1.5 cm long, 0.7–1 cm wide, densely hispidulous, rarely glabrous on the abaxial side. Petals 6–8, red, obovate, 2–4 cm long, 1.5–2 cm wide. Filaments entirely yellowish white to entirely purple-red. Disk fleshy, waved, ca. 1 mm high, yellow. Carpels 1–3, but often 2, ovoid, always tomentose, hairs green, yellow to purple-red; stigmas sessile, red, 1–1.5 mm wide. Follicles ovoid.

Flowering from middle April to late May; fruiting from August and September.

Chromosome number: 2n = 10.

Growing usually in steppes, meadows, open sandy dunes, bushes, or at the edges of forests, below 900 m altitude.

Armenia, Azerbaijan, Bulgaria, Georgia, Romania, Russia (the Caucasus), Serbia, Turkey (European part) and Ukraine on record.

Fig. 28.1 A large population in a meadow at 630 m, Temnolesskaya, Stavropol, Russia, 15 May 1999. [Voucher: *D. Y. Hong & S. L. Zhou* H99053 (A, CAS, K, MO, PE, US).]

Opposite:

Fig. 28.2 A rather large population in high thickets with scattered trees at 760 m, Mukhani, Kartli, Georgia, 5 May 1999. [Voucher: *D. Y. Hong & S. L. Zhou* H99043 (A, CAS, K, MO, PE, US).]

Fig. 28.3 Several individuals from the population shown in Fig. 28.2.

Fig. 28.4 An individual from the population shown in Fig. 28.2.

Fig. 28.5 A flower of another individual from the population shown in Fig. 28.2.

Fig. 28.6 Two types of leaves on the same individual at 630 m, Igoeti, Kartli, Georgia: the infertile shoot has broader leaf segments, 2 May 1999. [Voucher: *D. Y. Hong & S. L. Zhou* H99028 (A, CAS, K, MO, PE, US).]

Fig. 28.7 Polymorphism in floral parts in the population shown in Fig. 28.2: **a**, purple sepals and three green-yellow carpels; **b**, purple sepals, purple filaments and a single purple carpel; **c**, green (periphery purple) sepals, white filaments and two purple carpels.

Fig. 28.8 Fusiform roots from the population shown in Fig. 28.2.

29. ***Paeonia peregrina*** Mill., *Gard. dict. ed. 8*, No. 3 (1768).

Perennials. Lateral roots fusiform or tuberous. Stems 30–70 cm tall, glabrous. Lower leaves biternate; leaflets all or nearly all segmented, leaflets and/or leaf segments 17–45, lanceolate, oblanceolate or broad-linear, 3.5–11 cm long, 1.6–3.8 cm wide, always shallowly lobed, with 2–5 obtuse to acute lobes, often with bristles along veins above, usually glabrous beneath. Flowers terminal. Involucrate bracts 1–4, but usually 2 or 3. Sepals 3–5, mostly rounded at the apex. Corolla red or dark red, cup-shaped, petals 7–10, obovate, 4–5 cm long, 2.5–3 cm wide. Filaments yellow or red. Disk ca. 1 mm high, waved, white. Carpels 1–4, but mostly 2 or 3, tomentose, hairs ca. 2.5 mm long; stigmas sessile, yellow, 1.8–3.3 mm wide. Follicles 2.5–3.5 cm long. Seeds ovoid-oblong, black, lucid, 8–10 mm long, 5–6 mm wide.
Flowering from late May to June; fruiting in August.
Chromosome number: 2n = 20, but Sopova (1971) reported both 2n = 10 and 2n = 20 from Macedonia.
Growing mostly in deciduous forests, pine forests, or mixed forests, less frequently in grasses, in calcarious soils, below 1,500 m in altitude.
Distributed in Albania, Bulgaria, Greece, Italy, Macedonia, Moldova, Romania, Serbia and Turkey.

29.1
29.3
29.2

Opposite:

Fig. 29.1 A population at the edges of *Pinus nigra* forest with abundant individuals at 1,255 m, above Baglica Village, Nallihan, Ankara Province, Turkey, 11 May 2002. [Voucher: *D. Y. Hong et al.* H02201 (A, BM, CAS, K, MO, PE).]

Fig. 29.2 An individual near Selalmaz Village at 1,060 m, between Daday and Eflani, Kastamonu, Turkey, 22 May 2002. [Voucher: *D. Y. Hong et al.* H02223 (A, BM, CAS, K, MO, PE).]

Fig. 29.3 A flower in the same population as the plant shown in Fig. 29.2.

Above:

Fig. 29.4 Another flower in the same population as the plant shown in Fig. 29.2, showing a petaloid sepal, red filaments, a yellow disk, and yellowish ovaries and stigmas.

Fig. 29.5 An individual from the population shown in Fig. 29.1 but introduced into the Beijing Botanic Garden, Chinese Academy of Sciences, April 2003.

Fig. 29.6 The same flower as that shown in Fig. 29.5, back view.

Fig. 29.7 Fusiform roots from the population shown in Fig. 29.1.

30. ***Paeonia saueri*** D. Y. Hong, X. Q. Wang & D. M. Zhang, *Taxon* 53 (1): 88, figs 3 & 4 (2004).

Perennials. Lateral roots tuberous, tubers fusiform. Stems 45–65 cm tall, glabrous. Lower leaves biternate, with some leaflets segmented, leaflets and/or leaf segments 19–45, all entire or very few lobed, elliptic or narrow-elliptic, cuneate at the base, acute at the apex, 3.3–11 cm long, 1–4.2 cm wide, with bristles along veins above, usually sparsely hispidulous beneath. Flowers terminal. Involucrate bracts 2–3, leaf-like. Sepals 3–5, glabrous, all rounded but sometimes one caudate at the apex. Petals 7–10, red, obovate, 5–5.5 cm long, 3.2–4 cm wide. Disk fleshy, slightly waved or incised, ca. 1 mm high. Carpels 1–6 but mostly 3 or 2, whitish tomentose; stigmas sessile, red, about 2 mm wide. Young follicles ovoid, 2.8–3 cm long, 1.9 cm in diameter.
Flowering in April and May.
Chromosome number: 2n = 20.
Growing mostly near mountain summits, in deciduous forests, at the edges of forests or in clearings at an altitude of 400–1,220 m.
Confined to NE Greece and S Albania.

Fig. 30.1 A population in blossom at the edges of *Fagus sylvestris* forest at 910 m, Mt Pangeon, Kavala, Greece, 27 May 1985. Photograph by Dr W. Sauer.

Opposite:

Fig. 30.2 A population just over blossom near the population shown in Fig. 30.1, 29 May 2002. [Voucher: *D. Y. Hong, D. M. Zhang & X. Q. Wang* H02227 (A, BM, CAS, K, MO, PE, UPA).]

Fig. 30.3 An individual in flower in the population shown in Fig. 30.2.

Fig. 30.4 Another individual with purple leaves in the population shown in Fig. 30.2.

Fig. 30.5 (a & b) An individual from the population shown in Fig. 30.2, introduced into the Beijing Botanic Garden, Chinese Academy of Sciences, May 2003.

Fig. 30.6 Polymorphism of floral parts in the population shown in Fig. 30.2: **a**, a yellow disk, purple filaments, two tomentose carpels with a pinkish stigma; **b**, a red disk and two sparsely hairy carpels with a red stigma; **c**, purple sepals, white filaments, and three purplish carpels with a purple stigma; **d**, four tomentose carpels with a red stigma; **e**, five tomentose carpels with a purple stigma.

Fig. 30.7 Fusiform roots from the population shown in Fig. 30.2.

30.2
30.5a
30.5b
30.3
30.4
30.6a
30.6b
30.6c
30.7
30.6d
30.6e

31. *Paeonia arietina* G. Anderson, *Trans. Linn. Soc. London* 12: 275 (1818).

Paeonia mascula (L.) Mill. subsp. *arietina* (G. Anderson) Cullen & Heywood.

Perennials. Tap roots columnar; lateral roots always tuberous, sometimes tubers beaded. Stems 30–70 cm tall, usually entirely hirsute. Lower leaves biternate; leaflets usually with some segmented or shallowly divided; leaflets plus leaf segments 11–32 but mostly 13–23, elliptic, oblong or ovate-lanceolate, acute at the apex, 5–12 cm long, 3–6 cm wide, glabrous or villose along major veins above, but mostly densely, rarely sparsely villose beneath. Flowers terminal. Involucrate bracts 1–3, leaf-like. Sepals 3–5, all rounded at the apex, villose outside, green but purple at the periphery. Petals 6–9, rose or red, obovate. Disk waved, glabrous but sometimes hirsute. Carpels 1–5 but mostly 2 or 3, yellow tomentose, hairs 2.5 mm long; styles very short or absent; stigmas red, ca. 1 mm wide. Follicles oblong-ellipsoid, 2–3 cm long.

Flowering in May and June; fruiting in August and September.

Chromosome number: 2n = 20.

Growing usually in open oak or conifer forests, or clearings of forests, at altitudes from 300 to 2,100 m. Distributed in Albania, Bosnia-Herzegovina, Croatia, Italy (Emilia), Romania and Turkey.

Fig. 31.1 A plant from a population in a *Pinus nigra* forest at 1,000 m, between Ayazma and Kobakli, Mt Ida, Balikesir, Turkey, 13 May 2002. [Voucher: *D. Y. Hong et al.* H02204 (A, CAS, K, MO, PE, UPA).]

Fig. 31.2 A flower from the same population as the plant shown in Fig. 31.1.

Fig. 31.3 Back view of the flower shown in Fig. 31.2.

Fig. 31.4 In clearings, the population that includes the plant shown in Fig. 31.1 has abundant individuals with rose flowers.

Fig. 31.5 A rose-red flower from the population shown in Fig. 31.4.

Fig. 31.6 Even more individuals of the population shown in Fig. 31.4 are found on a slope.

Fig. 31.7 A flower from the population shown in Fig. 31.4 with purple filaments and two tomentose carpels.

Fig. 31.8 Another flower from the population shown in Fig. 31.4 with a pinkish disk, yellowish filaments, and two tomentose carpels.

Fig. 31.9 A population in a low secondary oak forest at 1,350 m, between Alucra and Sebinkarahisar, Giresum, Turkey, 19 May 2002. [Voucher: *D. Y. Hong et al.* H02218 (PE, UPA).]

Fig. 31.10 Fusiform roots from the population shown in Fig. 31.4

Fig. 31.11 A shoot from a broken root: evidence of cloning, in a pasture with scattered shrubs at 1,700 m, Zara, Sivas, Turkey, 19 May 2002. [Voucher: *D. Y. Hong et al.* H02216 (A, CAS, K, MO, PE, UPA).]

32. ***Paeonia parnassica*** Tzanoud., *Cytotax. Stud. Gen. Paeonia* L. *in Greece* 43 (1977).

Perennial herbs. Tap roots columnar; lateral roots tuberous or fusiform, sometimes tandem-fusiform. Stems sparsely to densely hirsute. Lower leaves biternate; leaflets 9, rarely 8, usually one or several segmented; leaflets and/or leaf segments 9–15, very rarely up to 25, ovate, oblong or elliptic, nearly rounded to acute at the apex, 4.5–12 cm long, 1.5–7 cm wide, glabrous above, densely or sparsely villose beneath. Flowers terminal. Involucrate bracts 1–3, leaf-like. Sepals 3 or 4, with one caudate and the rest rounded at the apex, nearly orbicular, densely villose on the abaxial side. Petals 6–8, dark purple, oblong or obovate, 4.5–6 cm long, 3–4.5 cm wide. Filaments purple; anthers purple. Disk 1–1.5 mm high, waved, tomentose. Carpels 1–3, but mostly 2, hairs 2 mm long; stigmas sessile, red, ca. 2 mm wide. Follicles columnar-ellipsoid, 1.4–2.2 cm long, 0.5–0.7 cm in diameter, yellowish tomentose.

Flowering from late May to early June.

Chromosome number: 2n = 20.

Growing at the edges and openings of *Abies* forests, or in sparse *Abies* forests on limestones at an altitude of 1,100–1,500 m.

Confined to the mountains Parnassos and Elikonas of Greece.

32.1 32.2 32.3

Opposite:

Fig. 32.1 A population on the edges of *Abies* forest at 1,210–1,250 m, Mt Parnassos, Greece, 25 May 2002. [Voucher: *D. Y. Hong et al.* H02224 (A, BM, CAS, K, MO, PE).]

Fig. 32.2 An individual from the population shown in Fig. 32.1.

Fig. 32.3 A flower of the plant shown in Fig. 32.2.

This page:

Fig. 32.4 A flower from the population shown in Fig. 32.1: back view showing the abaxial side of the entirely villous leaves, involucrate bracts and sepals.

Fig. 32.5 Polymorphism of floral parts in the population shown in Fig. 32.1: **a**, pinkish receptacle, yellow disk, dark purple filaments and two carpels; **b**, red receptacle, purple disk and two carpels; **c**, orange receptacle, red disk and three carpels.

Fig. 32.6 Fusiform roots from the population shown in Fig. 32.1.

Fig. 32.7 A prostrate, tandem-fusiform root from the population shown in Fig. 32.1.

33. *Paeonia officinalis* L., *Sp. pl.* 530 (1753).

Perennials. Roots tuberous-thickened. Caudex short, less than 10 cm long. Stems mostly hirsute. Lower leaves biternate; leaflets 9, usually segmented; leaflets plus leaf segments 11–130, linear-elliptic to elliptic, or oblanceolate, cuneate at the base, acuminate or acute at the apex, lobed or entire, 3–12 cm long, 1–4.5 cm wide, glabrous above, villose, rarely glabrous beneath. Flowers terminal. Involucrate bracts 1–2, leaf-like. Sepals 3–5, deltoid-orbicular to orbicular, mostly rounded at the apex, hispidulous or glabrous on the abaxial side. Petals 5–8, pale violet-red or purple-red. Disk up to 1 mm high, flat or waved, red. Carpels 1–5, but mostly 2 or 3, tomentose or glabrous, hairs 1.5–2 mm long, yellow, brown or pink; stigmas sessile, red, ca. 1.5 mm wide. Follicles long-ovoid when young. Chromosome number: 2n = 20.
Relatively widely distributed from the Iberian Peninsula to the Balkans via France, Italy and Switzerland. Five subspecies, of which four are represented here.

1a. Carpels glabrous, occasionally sparsely hirsute; sepals mostly glabrous; leaflets or leaf segments usually obtuse at the apex, often lobed; stems mostly glabrous . . . . . . 33e. subsp. *microcarpa*
1b. Carpels tomentose; sepals usually pubescent; leaflets or leaf segments usually acuminate at apex, entire, rarely lobed; stems hirsute or glabrous
2a. Leaflets plus leaf segments 11–24 in number, 2–4.5 cm wide, glabrous or sparsely villose beneath; sepals glabrous or sparsely pubescent . . . . 33b. subsp. *banatica*
2b. Leaflets plus leaf segments 19–130 in number, 1–3 cm wide, always villose beneath; sepals always pubescent
3a. Leaflets plus leaf segments 35–130 in number, always with some lobed, 1–2 cm wide. . . . . 33c. subsp. *huthii*
3b. Leaflets plus leaf segments 19–45 in number, all entire or occasionally very few lobed, 1–3 cm wide
4a. Leaves villose-floccose beneath, hairs flattened at base . . . . 33d. subsp. *italica*
4b. Leaves hairy but never villose-floccose beneath, hairs cylindrical . . 33a. subsp. *officinalis*

33a. *Paeonia officinalis* L. subsp. *officinalis* (Figs 33.1–33.5)

Paeonia peregrina Mill. var. *officinalis* (Retz.) Huth.

Flowering from late April to early June; fruiting in August and September.
Growing in open pine or oak woods, thickets or mountain meadows in limestone areas, at an altitude of 500–2,000 m.
Croatia, N Italy, Slovenia and S Switzerland on distribution record.

Opposite:

Fig. 33.1 A large population in a mountain meadow at 1,350–1,600 m, Mt Generoso, Lugano, Switzerland, 22 June 2001. [Voucher: *D. Y. Hong, X. Q. Wang & Y. M. Yuan* H01029 (A, BM, CAS, K, MO, P, PE).]

Fig. 33.2 An individual in the population shown in Fig. 33.1.

Fig. 33.3 A flower of another individual in the in the population shown in Fig. 33.1.

Fig. 33.4 A third individual in the population shown in Fig. 33.1, showing the shape of the leaves and flowers with two or three carpels.

Fig. 33.5 Fusiform roots from the population shown in Fig. 33.1.

33.1
33.2
33.3
33.4
33.5

33b. ***Paeonia officinalis*** L. subsp. ***banatica*** (Rochel) Soó, *Növényföldrajz*, 146 (1945).
Paeonia banatica Rochel. (Figs 33.6–33.8)

Flowering from April to late May; fruiting in August.
Growing in thickets or open woods of sand soils below 1,000 m in altitude.
Confined to the Balkans (Bosnia-Herzegovina, S Hungary, SW Romania and Serbia).

33c. ***Paeonia officinalis*** L. subsp. ***huthii*** Soldano, *Att. Soc. Ital. Sci. Nat. Museo Civ. Stor. Nat. Milano* 133 (10): 114 (1993). (Figs 33.9–33.13)
Paeonia officinalis L. subsp. *villosa* (Huth) Cullen & Heywood; *P. humilis* Retz. var. *villosa* (Huth) Stern.

Flowering from May to early June; fruiting in August and September.
Growing in open woods or in pastures with scattered trees, in limestone areas, at altitudes from 900 to 1,700 m.
Found in NW Italy, and SE and S France.

33d. ***Paeonia officinalis*** subsp. ***italica*** Passalacqua & Bernardo in Webbia, 59(2): 250 (2004).

The subspecies very much resembles subsp.*officinalis*, but differs from it in having hairs on the abaxial side of leaves flattened, not cylindrical at the base.
Confined to central Italy, Croatia and N Albania.

33e. ***Paeonia officinalis*** L. subsp. ***microcarpa*** (Boiss. & Reut.) Nym., *Consp. fl. eur.* 1: 22 (1878). (Figs 33.14–33.22)
Paeonia microcarpa Boiss. & Reut.; *P. humilis* Retz.; *P. officinalis* subsp. *humilis* (Retz.) Cullen & Heywood.

Flowering from April to May; fruiting in August and September.
Growing in pine woods or in thickets in limestone areas, at an altitude of 400–2,050 m.
Portugal, Spain and SW France on distribution record.

33.6 33.7

Opposite:

Fig. 33.6 A plant with two follicles and mature black seeds in a small population in a *Populus–Quercus–Pinus* forest, Flamunda, Deliblat, Banat region, Serbia, 17 August 2003. [Voucher: *D. Y. Hong, O. Vasic & V. Stojšić* H03020 (PE).]

Fig. 33.7 A plant grown from a section of old root in the population featured in Fig. 33.6.

Above:

Fig. 33.8 A horizontally running fusiform root from the population featured in Fig. 33.6.

Fig. 33.9 A rather large population in a sparse *Quercus–Fagus* wood at 1,130 m, N of Grasse, Alpes Maritimes, France, 26 May 2001. [Voucher: *D. Y. Hong, X. Q. Wang & A. Fridlender* H01009 (A, BM, CAS, K, MO, PE).]

Fig. 33.10 An individual with rose flowers in the population shown in Fig. 33.9.

Fig. 33.11 Polymorphism of floral parts in the population shown in Fig. 33.9: **a**, green sepals with purple periphery and a single carpel; **b**, green sepals and two carpels; **c**, purple sepals and two carpels.

Fig. 33.12 An individual with red flowers in a pasture with scattered trees of *Pinus sylvestris* at 1,130 m, above Castellar, Monton, Nice, France, 27 May 2001. [Voucher: *D. Y. Hong, X. Q. Wang & A. Fridlender* H01012 (A, BM, CAS, K, MO, PE).]

Fig. 33.13 A population at the edges of thickets and in thickets at 600–700 m, Montpellier, La Serane, France, 5 May 2001. Photography by Dr A. Fridlender. [Voucher: *A. Fridlender* H01007 (A, CAS, K, MO, PE).]

Fig. 33.14 A population on a stony hill at 550–650 m, Tour Madeloc, France–Spain frontier, France, 2 May 2001. Photograph by Dr A. Fridlender. [Voucher: *A. Fridlender* H01001 (K, MO, PE).]

Fig. 33.15 An individual in thickets at 360–600 m, C. de Banyuls, France–Spain frontier, France, 2 May 2001, by Dr A. Fridlender. [Voucher: *A. Fridlender* H01005 (A, K, MO, PE).]

Fig. 33.16 Another individual in the same population as that shown in Fig. 33.15, at the edges of thickets. Photograph by Dr A. Fridlender.

Fig. 33.17 A flower in the same population as the plant shown in Fig. 33.15. Photograph by Dr A. Fridlender.

Fig. 33.18 Polymorphism of floral parts in the population shown in Fig. 33.14: **a**, red disk, white filaments and two glabrous carpels; **b**, white disk, purple filaments and single glabrous carpel.

Fig. 33.19 Polymorphism of floral parts in a population in a *Pinus sylvestris* forest at 1,200 m, Hoyocasero, Avila, Spain, 10 August 2003. The photographed plants were introduced into the Beijing Botanic Garden, Chinese Academy of Sciences, April 2004: **a**, three carpels; **b**, four carpels. [Voucher: *D. Y. Hong & P. Vogas* H03016 (MO, PE).]

Fig. 33.20 A flower from the same population as the flowers shown in Fig. 33.19, back view, showing the glabrous abaxial side of the sepals and the villous abaxial side of the leaves.

Fig. 33.21 Two ripe follicles with blue-black seeds, from the same population as the flowers shown in Fig. 33.19.

Fig. 33.22 Fusiform roots from the same population as the flowers shown in Fig. 33.19.

INDEX TO BOTANICAL NAMES AND SYNONYMS

Accepted names in **bold**, synonyms in roman.